FUNDAMENTALS OF MATH
BOOK 2
ALGEBRA 1

2nd Edition

By
Jerry Ortner

AuthorHouse™
1663 Liberty Drive
Bloomington, IN 47403
www.authorhouse.com
Phone: 1 (800) 839-8640

Published by AuthorHouse   06/29/2015

ISBN: 978-1-4520-7000-1 (sc)
ISBN: 978-1-4520-6947-0 (e)

Print information available on the last page.

Any people depicted in stock imagery provided by Thinkstock are models,
and such images are being used for illustrative purposes only.
Certain stock imagery © Thinkstock.

This book is printed on acid-free paper.

Because of the dynamic nature of the Internet, any web addresses or links contained in this book may have changed
since publication and may no longer be valid. The views expressed in this work are solely those of the author and do
not necessarily reflect the views of the publisher, and the publisher hereby disclaims any responsibility for them.

TABLE OF CONTENTS

LESSON 1 – Algebraic Operations of Signed Numbers

When adding signed numbers, if the signs are all the same, keep that sign and add them. If the signs are different, take the sign of the largest number, and then subtract. When multiplying or dividing signed numbers and they have the same sign, the answer is positive. When the signs differ, the answer is negative.

Example 1:
$$(+6) + (+3) = +9$$

Example 2:
$$(-7) + (-3) = -10$$

Example 3:
$$(-6) + (+4) = -2$$

Example 4:
$$(-8) + (+9) = +1$$

Example 5:
$$(-8) + (-6) + (+9) =$$
$$(-14) + (+9) = -5$$

Example 6:
$$(-2)(-4) = 8$$

Example 7:
$$(-3)(5) = -15$$

Example 8:
$$(5)(4) = 20$$

Example 9:
$$(7)(-5) = -35$$

Example 10:
$$\frac{20}{-4} = -5$$

Example 11:
$$\frac{-32}{-8} = 4$$

Example 12:
$$\frac{-64}{16} = -4$$

Example 13:
$$\frac{48}{12} = 4$$

Example 14:
$$(+8) + (+9) + (+7) + (-3) + (-6) =$$
$$(+17) + (+7) + (-3) + (-6) =$$
$$(+24) + (-3) + (-6) =$$
$$(+21) + (-6) = +15$$

Example 15:
$$(-2) + (+3) + (+7) + (-5) =$$
$$(+1) + (+7) + (-5) =$$
$$(+8) + (-5) = +3$$

Example 16:
$$(-2) + (+3) + (-4) + (+5) =$$
$$(+1) + (-4) + (+5) =$$
$$(-3) + (+5) = +2$$

Practice on algebraic addition. Simplify the following.

1. $(-3) + (-6) + (+3) + (+8)$

2. $(-5) + (-8) + (+2) + (-9)$

3. $(-8) + (-3) + (+6) + (-7)$

4. $(-1) + (-4) + (+9) + (-2)$

5. $(-4) + (+6) + (-1) + (+4)$

6. $(-3) + (+2) - (+9)$

7. $(-2) + (+9) - (+2)$

8. $(-9) + (+7) - (+6)$

9. $(-8) + (+3) - (-4)$

10. $(-1) + (-2) + (+5) + (+8)$

11. $(-7) + (+4) + (+6) + (+7)$

12. $(-5) + (+6) + (+7) + (-6)$

13. $(-8) + (-7) + (+8) + (+4)$

14. $(-4) + (+8) + (-1) + (-2)$

15. $(-6) + (-9) + (+4) + (+1)$

16. $(-8) + (+2) - (+8)$

17. $(-3) + (+3) - (-2)$

18. $(-6) - (+6) - (-9)$

19. $(-4) + (+8) - (-3)$

20. $(-2) - (+9) - (+7)$

21. $(-5) - (+7) - (+1)$

22. $(-7) - (+5) + (+6)$

23. $(-1) + (+6) + (+1) + (+4)$

24. $(-8) + (+4) + (+1) + (+5)$

25. $(-7) + (+2) + (-5) + (-2)$

26. $(-3) + (+7) + (+3)$

27. a. $-1(-4)$
     b. $-9(+3)$
     c. $\dfrac{-16}{-4}$
     d. $\dfrac{25}{-5}$

28. a. $-6(-7)$
     b. $-4(+7)$
     c. $\dfrac{-25}{-5}$
     d. $\dfrac{12}{-3}$

29. a. $14(-80)$
     b. $11(+50)$
     c. $\dfrac{-170}{-17}$
     d. $\dfrac{450}{-9}$

30. a. $19(-30)$
     b. $14(+70)$
     c. $\dfrac{-144}{-16}$
     d. $\dfrac{490}{-7}$

31. a. $16(-80)$
     b. $15(+20)$
     c. $\dfrac{-104}{-13}$
     d. $\dfrac{100}{-5}$

32. a. $-8(-9)$
     b. $-6(+4)$
     c. $\dfrac{-20}{-5}$
     d. $\dfrac{9}{-3}$

33. a. $16(-40)$
     b. $17(+50)$
     c. $\dfrac{-99}{-11}$
     d. $\dfrac{150}{-5}$

34. a. $-5(-4)$
     b. $-5(+2)$
     c. $\dfrac{-12}{-3}$
     d. $\dfrac{15}{-3}$

35. a. $-3(-8)$
     b. $-2(+8)$
     c. $\dfrac{-12}{-4}$
     c. $\dfrac{8}{-2}$

36. a. $(-5)9$
     b. $(4)(-13)$
     c. $(-16)(-9)$
     d. $(8)(-3)(-4)$

Select the best answer for each operation.

37. −3(−20)
    a. −60   b. 60
    c. 70   d. −70

38. −2(−22)
    a. 44   b. −44
    c. 34   d. −34

39. $\dfrac{48}{-8}$
    a. −5   b. −6
    c. 6   d. 5

40. $\dfrac{63}{-9}$
    a. −6   b. 7
    c. −7   d. 6

41. $\dfrac{36}{-4}$
    a. −9   b. 10
    c. −10   d. 9

42. −9(−56)
    a. −494   b. 494
    c. −504   d. 504

43. −5(−28)
    a. −150   b. 150
    c. 140   d. −140

44. −4(−37)
    a. 138   b. −138
    c. 148   d. −148

45. $\dfrac{12}{-3}$
    a. −4   b. 5
    c. 4   d. −5

46. $\dfrac{20}{-4}$
    a. −6   b. 6
    c. 5   d. −5

47. −8(−40)
    a. 330   b. 320
    c. −330   d. −320

48. $\dfrac{24}{-3}$
    a. 8   b. −8
    c. −12   d. 12

49. $\dfrac{-160}{-20}$
    a. 2   b. 4
    c. 6   d. 8

50. (5)(−3)(−2)
    a. −30   b. 30
    c. 32   d. −32

LESSON 2 – Order of Operations
PEMDAS

Step 1:   Remove parentheses and exponents from left to right.
Step 2:   Multiply or divide moving from left to right.
Step 3:   Add or subtract moving from left to right.

Example 1:
$6 \cdot 5 + 4(3) - 8 =$
$30 + 12 - 8 =$
$42 - 8 = 34$

Example 2:
$5(4) + 3(2) - 6 =$
$20 + 6 - 6 =$
$26 - 6 = 20$

Example 3:
$4 - 2 + 8(5) =$
$4 - 2 + 40 =$
$2 + 40 = 42$

Example 4:
$(5 - 3) \cdot 6 + 4(-2 + 5) =$
$2 \cdot 6 + 4(3) =$
$12 + 12 = 24$

Example 5:
$(2^3 - 3) + 5^2 \cdot 3 =$
$(8 - 3) + 25 \cdot 3 =$
$5 + 25 \cdot 3 =$
$5 + 75 = 80$

Example 6:
$[6 \cdot (-2)] + 5(4 - 1) + 6 + 12 =$
$-12 + 5(3) + 6 + 12 =$
$-12 + 15 + 6 + 12 =$
$3 + 6 + 12 =$
$9 + 12 = 21$

Example 7:
$-(4 - 6) + 5[4 + (-7)] + 20 =$
$-(-2) + 5(-3) + 20 =$
$2 - 15 + 20 =$
$-13 + 20 = 7$

Example 8:
$5^2 + (3 \times 7) - 4 =$
$25 + 21 - 4 =$
$46 - 4 = 42$

Example 9:
$10 + 2^2 + 4 =$
$10 + 4 + 4 =$
$14 + 4 = 18$

Example 10:
$10^2 - 7^2 + 5 =$
$100 - 49 + 5 =$
$51 + 5 = 56$

Practice problems on order of operations. Simplify or evaluate the following.

1. $2 \cdot 9 - 5(4) + 7$
     a. 5   b. 45
     c. 88   d. 39

2. $5 + 3 \cdot 4 - 2$

3. $3 + 2 \cdot 5 - 4$

4. $2 + 3 \cdot 4 - 5$

8

5. $5 \cdot 9 - 2(-8) + 7$
   a. 68   b. –35
   c. 36   d. –273

6. $7 \times 2 - 6(-3) + 4$
   a. 88   b. 36
   c. 0   d. –28

7. $4 \times 6 - 5(-9) + 8$
   a. –4   b. –28
   c. 77   d. –13

8. $8 \cdot 5 - 7(2) + 3$
   a. 57   b. –29
   c. –80   d. 29

9. $3 \cdot 7 - 4(-6) + 9$
   a. 6   b. 27
   c. 54   d. –45

10. $2 + 5 \times 4 - 3$

11. $8 \cdot 9 - 6(5) + 4$
    a. 106   b. 216
    c. 124   d. 46

12. $4 \cdot 6 - 3(-7) + 2$
    a. –60   b. 47
    c. 5   d. –82

13. $2 \times 3 - 8(-9) + 5$
    a. 40   b. 95
    c. 83   d. –61

14. $5 \cdot 8 - 4(-6) + 7$
    a. –113   b. 20
    c. 23   d. 71

15. $-(-5) - 3(9 - 6)$
    a. 14   b. –28
    c. –40   d. –4

16. $-(-6) - 9(10 - 5)$
    a. –129   b. 51
    c. –89   d. –39

17. $-(-8) - 8(10 - 7)$

18. $6(-4 + 5)(-3 + 3) + 6$

19. $7(-9 + 4)(-1 + 7) + 2$

20. $4(-2 + 9)(-4 + 6) + 5$

21. $-(-9) - 7(8 - 2)$
    a. 51   b. –33
    c. –61   d. –49

22. $-2(-2) - 5(11 - 7)$
    a. –88   b. –60
    c. 22   d. –16

23. $-(-6) - 5(10 - 8)$

24. $-(-7) - 6(6 - 4)$
   a. -5 b. -33
   c. -53 d. 19

25. $-(-9) - 4(9 - 6)$
   a. -3 b. -33
   c. -51 d. 21

26. $-(-8) - 3(8 - 6)$

27. $-2(4 + 2)(9 + 4) + 8$

28. $-8(-2 + 4) - 3(4 - 3)$

29. $-4(-3 + 6)$
   a. -6 b. -12
   c. 18 d. 12

30. $44 - 2(5 + 3)$

31. $40 - 3(7 - 4)$

32. $46 - 4(6 + 2)$

33. $-4(-5 + 4)$
   a. 24 b. -4
   c. -16 d. 4

34. $2(-4 + 2)$
   a. 10 b. -4
   c. -6 d. 4

35. $-2(-7 + 5) - 3(2 - 8)$

36. $-6(4 + 3) - 7(-7 - 1)$

37. $-2(-6 + 7)(7 + 6) + 8$

38. $5(9 + 1)(4 + 7) + 2$

39. $8(-2 + 9) - 2(-6 - 9)$

40. $3(-4 + 3)$
   a. 3 b. -3
   c. 15 d. -9

41. $3(2 + 6) - 2(-7 - 2)$

42. $7(9 + 7) - 3(-3 - 9)$

43. $-2(-5 + 2)$
   a. 12   b. 6
   c. -8 d. -6

44. $3(-5 + 4) - 9(-7 - 6)$

45. $2 + 5 \cdot 3^2$
   a. 63   b. 47
   c. 227   d. 289

46. $6 + 4^2$

47. $8^2 - 5 - 7 \times 4$

48. $6^2 - 2 - 2 \times 6$

49. $8^2 - 4 - 5 \cdot 2$

50. $3^3 - 2^3 - 2 \cdot 3$

51. $9 + 1^3$

52. $8 + 2^3$

53. $1^2 - 1$

54. $4^3 - 3^3 - 3 \times 4$

55. $10^2 - 4 - 2 \times 9$

56. $8 + 4 \cdot 3^2$

57. $7^2 - 5 - 8 \times 3$

58. $6^3 - 3^3 - 3 \cdot 6$

59. $(3 + 2)(4)$

60. $\dfrac{4^3 + 2}{6}$

61. $(2^3 + 1^2)5$

62. $(5 - 3)(2^2)$

63. $(-3 - 4)(4)$

64. $9^2 - 3 - 6 \cdot 3^2$

65. $[(1 + 5)^2] \div 12$

66. $3^3 - 5^2 + 4^2$

11

LESSON 3 – Removal of Brackets/Parentheses

Example 1:
$$6 + (4 \cdot 3) - 5 =$$
$$6 + 12 - 5 =$$
$$18 - 5 = 13$$

Example 2:
$$4 + (9 \cdot 2) + 7 =$$
$$4 + 18 + 7 =$$
$$22 + 7 = 29$$

Example 3:
$$3(7 - 4) + 15 =$$
$$3(3) + 15 =$$
$$9 + 15 = 24$$

Example 4:
$$2(5 - 2)(1 + 3) - 8 =$$
$$2(3)(4) - 8 =$$
$$6(4) - 8 =$$
$$24 - 8 = 16$$

Example 5:
$$7(3 - 5) + 22 =$$
$$7(-2) + 22 =$$
$$-14 + 22 = 8$$

Example 6:
$$5(1 - 4)(3 - 4) + 9 =$$
$$5(-3)(-1) + 9 =$$
$$(-15)(-1) + 9 =$$
$$15 + 9 = 24$$

Example 7:
$$6 + \left(\frac{18}{9}\right) - 5 =$$
$$6 + 2 - 5 =$$
$$8 - 5 = 3$$

Example 8:
$$10 + \frac{32}{-4} + 6 =$$
$$10 + (-8) + 6 =$$
$$2 + 6 = 8$$

Example 9:
$$15 + \frac{24}{-6} - 20 =$$
$$15 + (-4) - 20 =$$
$$11 - 20 = -9$$
or
$$15 + \frac{24}{-6} - 20 =$$
$$15 - 4 - 20 =$$
$$11 - 20 = -9$$

Example 10:
$$\frac{(4 - 2)(5 - 6) + 18}{(-7) + 9} =$$
$$\frac{2(-1) + 18}{2} =$$
$$\frac{-2 + 18}{2} =$$
$$\frac{16}{2} = 8$$

Solve the following.

1. $3(-2 + 5)$
    a. −1   b. 9
    c. 11   d. −9

2. $2(-5 + 6)$
    a. −4   b. 2
    c. 16   d. −2

3. $50 - 4(6 + 2)$

4. $43 - 3(7 - 4)$

5. $4(-2 + 5)$
       a. 13   b. 12
       c. −12   d. −3

6. $-2(-5 + 2)$
       a. 12   b. −6
       c. −8   d. 6

7. $7(-6 + 5) - 7(-5 - 7)$

8. $-9(3 + 1) - 5(1 - 3)$

9. $4(7 + 6)(-5 + 3) + 5$

10. $7(6 + 4)(2 + 4) + 7$

11. $2 + \dfrac{-16}{8} + 3$

12. $\dfrac{(6 + 1) + (-5 - 5)}{-8 + (-9)}$

13. $\dfrac{(8 - 4) + (6 + 5)}{8 - (-3)}$

14. $-3 + \dfrac{6}{-2} - 7$

15. $4(-2 + 4)(8 + 2) + 8$

16. $-3(6 + 7) - 5(-8 - 4)$

17. $-4(-2 + 1)$

18. $3(-3 + 5)$

19. $40 - 2(5 - 2)$

20. $40 - 2(6 - 4)$

21. $-2(-2 + 1)$

22. $-4(9 + 4) - 7(-3 - 1)$

23. $-8(-2 + 4) - 3(4 - 3)$

24. $5(3^2 - 2^3) + 15$

25. $16(5^2 - 4^2) + 16$

26. $[5(4 + 3)] - 5^2$

LESSON 4 – Evaluate a Square Root, Cube Root, and Absolute Value

When working with absolute value problems, perform the mathematics inside the absolute value symbols (| |) first, then extract the positive answer to continue. These examples will help.

Example 1:
$$\sqrt{16} = 4$$

Example 2:
$$\sqrt{25} = 5$$

Example 3:
$$\sqrt{121} = 11$$

Example 4:
$$\sqrt[3]{8} = 2$$

Example 5:
$$\sqrt[3]{64} = 4$$

Example 6:
$$3^3 + \sqrt[3]{125} =$$
$$27 + 5 = 32$$

Example 7:
$$|3| = 3$$

Example 8:
$$|-4| = 4$$

Example 9:
$$6 + |-3 + 2| =$$
$$6 + |-1| =$$
$$6 + 1 = 7$$

Example 10:
$$5 + |2 - 8| =$$
$$5 + |-6| =$$
$$5 + 6 = 11$$

Example 11:
$$3 - 8 + (6 + 4) - |10| =$$
$$3 - 8 + 10 - 10 =$$
$$-5 + 10 - 10 =$$
$$5 - 10 = -5$$
   or
$$3 - 8 + (6 + 4) - |10| =$$
$$3 - 8 + 10 - 10 =$$
$$-5 + 0 = -5$$

Example 12:
$$4 + |2 - 5| - |5 - 3| =$$
$$4 + |-3| - |2| =$$
$$4 + 3 - 2 =$$
$$7 - 2 = 5$$

Example 13:
$$|12 - 15| = |-3| = 3$$

Example 14:
$$5 + |1 - 4| =$$
$$5 + |-3| =$$
$$5 + 3 = 8$$

Example 15:
$$10 + |2 - 8| =$$
$$10 + |-6| =$$
$$10 + 6 = 16$$

Example 16:
$$|4 - 8| - |8 - 4| =$$
$$|-4| - |4| =$$
$$4 - 4 = 0$$

Practice on evaluating square and cube roots. Use the square root key on your calculator to find the square root. Then solve the problem using PEMDAS (order of operations). Simplify the following.

1. $\sqrt{9}$

2. $\sqrt{36}$

3. $\sqrt{49}$

4. $\sqrt{144}$

5. $\sqrt{100}$

6. $\sqrt{81}$

7. $\sqrt{25} + \sqrt{36}$

8. $(\sqrt{4})(\sqrt{9})$

9. $\sqrt{225}$

10. $\sqrt{169}$

11. $(\sqrt{144})(\sqrt{4})$

12. $\sqrt{64}$

13. $\sqrt{121} + \sqrt{100}$

14. $\sqrt{169}$

15. $(\sqrt{4})(\sqrt{9})(\sqrt{16})$

16. $\sqrt{100} - \sqrt{36}$

17. $\sqrt{16} + \sqrt{9}$

18. $(\sqrt{25})(\sqrt{49})$

19. $2(\sqrt{4})$

20. $3(\sqrt{4}) + \sqrt{16}$

21. $\sqrt{36} - \sqrt{16}$

22. $\sqrt{49}(\sqrt{9})$

23. $\sqrt{25} + \sqrt{25} + \sqrt{25}$

24. $(\sqrt{64})(\sqrt{81})$

25. $(\sqrt{100})(\sqrt{64})$

26. $\sqrt[3]{27}$

27. $\sqrt[3]{125}$

28. $4^2 + \sqrt[3]{27}$

29. $2^2 + \sqrt[3]{64}$

30. $\sqrt[3]{8} + 2^2$

31. $4^2 + \sqrt[3]{125}$

32. $3^3 + \sqrt[3]{27}$

33. $9\left(\sqrt[3]{8}\right)$

34. $\left(\sqrt[3]{1}\right)\left(\sqrt[3]{64}\right)\left(\sqrt[3]{27}\right)$

35. $\left(\sqrt[3]{125}\right)\left(\sqrt{25}\right)$

36. $4^3 - \sqrt[3]{64}$

37. $(18)(1^3) + \sqrt[3]{8}$

38. $6^2 + \left(\sqrt[3]{8}\right)\left(\sqrt[3]{27}\right)$

39. $5^2 - \left(\sqrt[3]{125}\right)\left(\sqrt[3]{8}\right)$

40. $\left(\sqrt{36} + \sqrt[3]{64}\right) - \sqrt{144}$

41. $\sqrt[3]{8}\left(\sqrt[3]{27} + \sqrt{4}\right)$

42. $|-4 + 4| - 5 - 3$

43. $|1 - 3| + 6 + 8$

44. $-|-7 + 4| - 7 + 7$

45. $|1 - 5| + 5 + 7$

46. $-7 + |-7 - 6| - 4$
    a. −4   b. 4
    c. −24   d. 2

47. $2 - |-3 + 4| + 7$
    a. −2   b. 8
    c. −8   d. 2

48. $3 + |7 + 2| - 9$
    a. −3   b. 7
    c. 3   d. −17

49. $-(-6) - (-1) - 3$

50. $7 - |-5 + 3| - 4$
   a. 5   b. 1
   c. –19   d. 19

51. $-9 + |2 - 4| + 4$
   a. –3   b. –7
   c. 7   d. –19

52. $4 - |3 - 3| - 8$
   a. 4   b. –4
   c. 18   d. –6

53. $-7 + |-4 + 3| + 5$
   a. –3   b. –19
   c. 19   d. –1

54. $-(-8) - |-11| + 5$

55. $-(-1) - (-5) - 3$

56. $10 - 6 - |-14| + |11 - 10 + 6|$

57. $12 - 14 - |-15| + |7 - 2 - 5|$

58. $4 - 7 - (-12) + |3 - 13 - 4|$

59. $7 - |4 + 6| - 7$
   a. 16   b. –10
   c. –16   d. 10

60. $|-10(4) + 33| - |16 - 2^3|$

LESSON 5 – Left to Right Inside Parentheses with Numerous Signs

Remember: PEMDAS – Parentheses, Exponents, Multiply/Divide, Add/Subtract
Note: The symbols | | represents absolute value.

Example 1:
$$(-4) + |-\{-4\}| =$$
$$-4 + |+4| =$$
$$-4 + 4 = 0$$

Example 2:
$$(6) + |-10| - (5)[-(-(-4))] =$$
$$6 + 10 - 5[-4] =$$
$$6 + 10 - 5[-4] =$$
$$16 + 20 = 36$$

Example 3:
$$(-[-3]) + |5(-5)| =$$
$$(3) + |-25| =$$
$$3 + 25 = 28$$

Example 4:
$$|5 - 8| + [-(-[-8])] =$$
$$|-3| + [-8] =$$
$$3 + [-8] =$$
$$3 - 8 = -5$$

Example 5:
$$(-[-(-[-6])]) + |6 - [-10]| =$$
$$(+6) + |6 + 10| =$$
$$6 + |16| =$$
$$6 + 16 = 22$$

Example 6:
$$5 - [-(7)] + (-8)(1) =$$
$$5 + 7 + (-8) =$$
$$5 + 7 - 8 =$$
$$12 - 8 = 4$$

Example 7:
$$9 - 7 + [-(6)] - |-13| =$$
$$9 - 7 - 6 - 13 =$$
$$2 - 6 - 13 =$$
$$-4 - 13 = -17$$

Example 8:
$$|-6| - [-3(-2)] + (7)(-3) =$$
$$6 - [+6] + (-21) =$$
$$6 - 6 - 21 =$$
$$0 - 21 = -21$$

Example 9:
$$(-8) + [-(-4)] + (4)(-2)(-1) =$$
$$-8 + [+4] + 4(+2) =$$
$$-8 + 4 + 8 =$$
$$-4 + 8 = 4$$

Example 10:
$$[2] + |6 - 8| - (2)[-(-4)] =$$
$$2 + |-2| - (2)[+4] =$$
$$2 + 2 - 8 =$$
$$4 - 8 = -4$$

Example 11:
$$(-[6]) + (5)(-3) + |-4| =$$
$$(-6) + (-15) + 4 =$$
$$-6 - 15 + 4 =$$
$$-21 + 4 = -17$$

Example 12:
$$[5] + (7)(-2) - [-(-[6])] =$$
$$5 + (-14) - [+6] =$$
$$5 - 14 - 6 =$$
$$-9 - 6 = -15$$

Practice on left to right inside parentheses with lots of signs. Simplify the following.

1. −|−(−3)| − {−[−(−6)]}

2. −|−(−15)| − {−[−|−5|]}

3. 9 − 7 − (−11) − [−|−3|]

4. 6 − 4 − (−7) + |3 − 1 + 12|

5. 3 − 3 − |−13| + |5 − 10 + 14|

6. −(−6) − (−2) + (−6)

7. 3 − 8 − |−9| + |12 − 1 + 6|

8. 4 − 2 − (−4) − [−|−2|]

9. −[−(−2)] − {−[−(−14)]}

10. −(−1) − |−12| + 6
    a. 19   b. −5
    c. −7   d. 7

11. −[−(−2)] − {−[−(−13)]}

12. 1 − 13 − |−4| − [−(−10)]

13. −(−6) − (−14) + 13
    a. 7   b. −7
    c. 5   d. 33

14. −(−12) − (−13) − (−9)
    a. −10   b. −34
    c. 34   d. 16

15. −|−(− 4)| − {−[−|−6|]}

16. 6 − 8 − (−7) − [−|−11|]

17. 4 − 7 − (−5) − [−(−2)]

18. −[−(− 4)] − {−[−|−9|]}

19. −|−(−11)| − {−[−|−5|]}

20. −(−10) − (−9) + (−12)
    a. −31   b. −11
    c. 7   d. 31

21. −(−11) − (−10) + 7
    a. −14   b. 14
    c. 8   d. 28

22. −|−10| − |−5| + 2
    a. 7   b. 17
    c. 13   d. −13

23. −[−(−3)] − {−[−|−11|]}

24. −|−(−10)| − {−[−|−10|]}

19

25. $-|-(-2)| - \{-[-|-6|]\}$

26. $10 - 1 - (-5) - [-|-9|]$

27. $3 - 11 - |-14| + |2 - 13 - 10|$

28. $-(-7) - (-6) + (-14)$

29. $-7 + |-6 - 8| + 3$
      a. −18   b. 8
      c. 10   d. −12

30. $-(-2) - (-6) + (-13)$
      a. 21   b. −21
      c. −5   d. −17

31. $-(-8) - (-11) - 6$
      a. 13   b. 3
      c. −13   d. 25

32. $-(-5) - (-7) + 15$
      a. −3   b. 27
      c. 3   d. 13

33. $3 - 7 - (-15) + |12 - 6 - 11|$

LESSON 6 – Evaluating Exponents with a Negative Base and Signed Numbers

Example 1:
$$(-2)^2 =$$
$$(-2)(-2) = 4$$

Example 2:
$$(-2)^3 =$$
$$(-2)(-2)(-2) = -8$$

Example 3:
$$(-4)^2 =$$
$$(-4)(-4) = 16$$

Example 4:
$$5^2 - (-2)^3 =$$
$$25 - (-2)(-2)(-2) =$$
$$25 - (-8) =$$
$$25 + 8 = 33$$

Example 5:
$$(-3)^2 + 8^2 =$$
$$(-3)(-3) + (8)(8) =$$
$$9 + 64 = 73$$

Example 6:
$$-5^2 + 15 =$$
$$-(5^2) + 15 =$$
$$-(5)(5) + 15 =$$
$$-25 + 15 = -10$$

Example 7:
$$-3^3 + \sqrt[3]{64} =$$
$$-(3)(3)(3) + 4 =$$
$$-27 + 4 = -23$$

Example 8:
$$\left(\sqrt[3]{8}\right)(-2)^3 =$$
$$2(-2)(-2)(-2) =$$
$$2(-8) = -16$$

Example 9:
$$\left(\sqrt[3]{8}\right) + (-2)^3 =$$
$$2 + (-2)(-2)(-2) =$$
$$2 + (-8) =$$
$$2 - 8 = -6$$

Example 10:
$$-3^4 + \sqrt[3]{125}\left(\sqrt[3]{64}\right) =$$
$$-(3)(3)(3)(3) + 5(4) =$$
$$-81 + 20 = -61$$

Example 11:
$$(-1)^2 - (-3)^2 =$$
$$(-1)(-1) - (-3)(-3) =$$
$$1 - 9 = -8$$

Example 12:
$$-2^2 + 4^2 =$$
$$-(2)(2) + (4)(4) =$$
$$-4 + 16 = 12$$

Example 13:
$$(-2)^2 + 4^2 =$$
$$(-2)(-2) + (4)(4) =$$
$$4 + 16 = 20$$

Example 14:
$$-2^2 - 4^2 =$$
$$-(2)(2) - (4)(4) =$$
$$-4 - 16 = -20$$

Example 15:
$$(-2)^2 + (-4)^2 =$$
$$(-2)(-2) + (-4)(-4) =$$
$$4 + 16 = 20$$

Example 16:
$$-2^2 + (-4)^2 =$$
$$-(2)(2) + (-4)(-4) =$$
$$-4 + 16 = 12$$

Simplify these practice problems evaluating exponents with negative bases.

1. $(-3)^3$

2. $-(2)^4$

3. $-(3)^2$

4. $-2^3 - (-2)^2 - (-3)^3 + \sqrt[3]{64}$

5. $-2^4 - (-3)^2 - (-2)^4 + \sqrt[3]{8}$

6. $-3^3 - (-2)^4 - (-3)^2 + \sqrt[3]{27}$

7. $2^2 + \sqrt[3]{8}$

8. $3^3 + \sqrt[3]{27}$

9. $(-3)^3 + \sqrt[3]{27}$

10. $(-2)^3 + (-1)^4$

11. $4^2 + \sqrt[3]{64}$

12. $-3^2 - (-3)^2 - (-2)^4 + \sqrt[3]{27}$

13. $-2^2 - (-2)^3 - (-3)^2 + \sqrt[3]{8}$

14. $-2^4 - (-2)^2 - (-3)^3 + \sqrt[3]{64}$

15. $-3^2 - (-2)^2 - (-2)^4 + \sqrt[3]{27}$

16. $-2^4 - (-2)^3 - (-3)^2 + \sqrt[3]{8}$

17. $-3^2 - (-2)^2$

18. $-4^2 - (-3)^2$

19. $-2^4 - (-2)^2 - 2^2$

20. $-3^2 - (-3)^3 - 2^3$

21. $-3^3 - (-3)^2 - 2^4$

22. $-3^3 - (-2)^4$

23. $-(-4)^3 - (-3)^2$

24. $-3^3 - (-2)^3 - 2^2$

25. $-2^3 - (-3)^3 - 3^3$

26. $-(-4)^2 - (-2)^2$

27. $-4^2 - (-2)^2$

28. $-3^3 - (-2)^3 - 2^3$

LESSON 7 – Simple Mathematical Equations
Using Addition and Subtraction Rules

Example 1: $x + 5 = 8$

$$
\begin{array}{rcr}
x + 5 &=& 8 \\
-\ 5 & & -5 \\
\hline
x &=& 3
\end{array}
$$

Example 2: $16 + N = 35$

$$
\begin{array}{rcr}
16 + N &=& 35 \\
-16 & & -16 \\
\hline
N &=& 19
\end{array}
$$

Example 3: $a - 15 = 24$

$$
\begin{array}{rcr}
a - 15 &=& 24 \\
+\ 15 & & +15 \\
\hline
a &=& 39
\end{array}
$$

Example 4: $33 - N = 78$

$$
\begin{array}{rcr}
33 - N &=& 78 \\
-33 & & -33 \\
\hline
-\ N &=& 45 \\
N &=& -45
\end{array}
$$

Example 5: $35 - b = 14$

$$
\begin{array}{rcr}
35 - b &=& 14 \\
-35 & & -35 \\
\hline
-\ b &=& -21 \\
b &=& 21
\end{array}
$$

Example 6: $c - 14 = 32$

$$
\begin{array}{rcr}
c - 14 &=& 32 \\
+\ 14 & & +14 \\
\hline
c &=& 26
\end{array}
$$

Example 7: $9 + e = 26$

$$
\begin{array}{rcr}
9 + e &=& 26 \\
-9 & & -9 \\
\hline
e &=& 17
\end{array}
$$

Example 8: $f + 13 = -56$

$$
\begin{array}{rcr}
f + 13 &=& -56 \\
-\ 13 & & -13 \\
\hline
f &=& -69
\end{array}
$$

Practice problems on simple equations using addition and subtraction. Solve for the variable.

1. $N + 7 = 33$

2. $N + 6 = 40$

3. $n - 2 = 29$

4. $y - 4 = 7$

5. $j - 9 = 27$

6. $r - 9 = 15$

7. $t - 6 = 10$ 8. $c - 3 = 11$

9. $N - 8 = 20$ 10. $N - 8 = 18$

11. $g - 4 = 21$ 12. $j - 6 = 23$

13. $N + 9 = 36$ 14. $N + 8 = 39$

15. $N + 6 = 39$ 16. $3 + f = 11$

17. $9 + e = 15$ 18. $N + 6 = 43$

19. $c - 3 = 25$ 20. $v - 7 = 9$

21. $j - 4 = 18$ 22. $d - 6 = 14$

23. $N - 8 = 19$ 24. $N - 9 = 28$

25. $6 + x = -36$ 26. $p + 16 = -8$

27. $32 - Q = 57$ 28. $14 - M = -16$

29. $19 + a = 57$ 30. $N + 31 = -63$

31. $S + 37 = -58$ 32. $41 - S = -72$

33. $Z - 16 = 47$

LESSON 8 – Simple Mathematical Equations
Using the Division and Multiplication Rules

Example 1:
$$6x = 18$$
$$\frac{6x}{6} = \frac{18}{6}$$
$$x = 3$$

Example 2:
$$4 \cdot N = 36$$
$$\frac{4 \cdot N}{4} = \frac{36}{4}$$
$$N = 9$$

Example 3:
$$5a = 8 \cdot 10$$
$$5a = 80$$
$$\frac{5a}{5} = \frac{80}{5}$$
$$a = 16$$

Example 4:
$$b \div 6 = 8$$
$$\frac{b}{6} = 8$$
$$\frac{6 \cdot b}{6} = 8 \cdot 6$$
$$b = 48$$

Example 5:
$$7 \times n = 28$$
$$\frac{7 \times n}{7} = \frac{28}{7}$$
$$n = 4$$

Example 6:
$$4e = 8 \cdot 7$$
$$4e = 56$$
$$\frac{4e}{4} = \frac{56}{4}$$
$$e = 14$$

Example 7:
Solve for m: $3m = 36$
$$\frac{3m}{3} = \frac{36}{3}$$
$$m = 12$$

Example 8:
$$\frac{x}{4} = 13$$
$$\frac{4 \cdot x}{4} = 13 \cdot 4$$
$$x = 52$$

Example 9:
$$x \div 3 = 17$$
$$\frac{x}{3} = 17$$
$$\frac{3 \cdot x}{3} = 17 \cdot 3$$
$$x = 51$$

Example 10:
$$32 \div N = 8$$
$$\frac{N \cdot 32}{N} = 8 \cdot N$$
$$32 = 8 \cdot N$$
$$\frac{32}{8} = \frac{8 \cdot N}{8}$$
$$4 = N$$

Practice problems on simple equations using the division rule. Solve for the unknown.

1. $7h = 56$

2. $6s = 12$

3. $m \times 10 = 70$

4. $12 \times c = 36$

5. $8p = 120$

6. $4x = 32$

7. $4z = 32 \cdot 5$

8. $2m = 5 \cdot 2$

9. $11b = 33$

10. $4r = 36$

11. $7w = 21$

12. $7f = 4 \cdot 7$

13. $\dfrac{g}{6} = 3$

14. $\dfrac{m}{3} = 2$

15. $n \div 5 = 12$

16. $c \div 3 = 10$

17. $\dfrac{N}{5} = 7$

18. $\dfrac{N}{6} = 7$

19. $\dfrac{N}{3} = 5$

20. $j \div 8 = 10$

21. $\dfrac{t}{4} = 3$

22. $x \div 9 = 11$

23. $\dfrac{N}{5} = 3$

24. $\dfrac{u}{3} = 7$

25. $d \div 3 = 11$

26. $p \div 2 = 9$

27. $\dfrac{f}{8} = 10$

28. $\dfrac{c}{2} = 6$

29. $4u = 10 \cdot 16$

30. $6x = 7 \cdot 18$

LESSON 9 – Solving Equations with Several Rules

Example 1: $6x + 5 = 23$

    Solution:

$$
\begin{array}{rrrl}
6x & + \quad 5 & = \quad 23 & \\
\text{Step 1} \qquad & - \quad 5 & \quad - 5 & \text{Subtraction 5 from both sides.} \\
\hline
& 6x & = \quad 18 & \\
\text{Step 2} \qquad & \dfrac{6x}{6} & = \quad \dfrac{18}{6} & \text{Divide both sides by 6.} \\
& x & = \quad 3 &
\end{array}
$$

Example 2: $\dfrac{x}{3} - 4 = 8$

    Solution:

$$
\begin{array}{rrrl}
\dfrac{x}{3} & - \quad 4 & = \quad 8 & \\
\text{Step 1} \qquad & + \quad 4 & \quad + 4 & \text{Add 4 to both sides.} \\
\hline
\dfrac{x}{3} & = \quad 8 & & \\
\text{Step 2} \qquad \dfrac{3x}{3} & = \quad 12(3) & & \text{Multiply both sides by 3.} \\
x & = \quad 36 & &
\end{array}
$$

Example 3: $4x + 6x - 8 = 22 + 30$

    Solution:

$$
\begin{array}{rrrrl}
\text{Step 1} \qquad & 10x & - \quad 8 & = \quad 52 & \text{Combine like terms on both sides.} \\
\text{Step 2} \qquad & & + \quad 8 & = \quad + 8 & \text{Add 8 to both sides.} \\
\hline
& 10x & = \quad 60 & & \\
\text{Step 3} \qquad & \dfrac{10x}{10} & = \quad \dfrac{60}{10} & & \text{Divide both sides by 10.} \\
& x & = \quad 6 & &
\end{array}
$$

Example 4: $4x - 7 + 2 = 30 + 5 - 4x$

Solution:

Step 1	$4x$	$-$	5	$=$	35	$- 4x$	Combine like terms.
Step 2		$+$	5		$+ 5$		Add 5 to both sides.

$$4x = 40 - 4x$$

Step 3 $+ 4x$ $+ 4x$ Add 4x to both sides.

$$8x = 40$$

Step 4 $\dfrac{8x}{8} = \dfrac{40}{8}$ Divide both sides by 8.

$$x = 5$$

Practice solving equations with two or more rules.

1. $5x - 1 = 9$ 2. $3x + 2 = 17$

3. $4x + 7 = 23$ 4. $2x + 5 = 15$

5. $-6x + 2 - 3x + 2 = 3 + 2x + 4x - 4$

6. $-2x - 1 + 5x + 3 = 1 + 2x + 3x - 4$

7. $-3x - 1 - 2x - 3 = -4 - x + 2x + 4$

8. $3x + 1 + 5x + 5 = -4 - 4x + 3x + 2$

9. $4x + 9 = 17$ 10. $5x + 1 = 31$

11. $4x + 7 = 35$ 12. $4x + 7 = 47$

13. $10x + 1 = 21$ 14. $8x + 5 = 61$

15. $4x + 7 = 11$

LESSON 10 – Variables on Both Sides of the Equation

Remember to get all unknowns to one side of the equation (preferably to the left side) and all constants to the other side before solving.

Example 1: $6x - 4x + 8 = x + 9 + 2$

Solution:

| $2x$ | $+$ | 8 | $=$ | x | $+$ | 11 | Combine like terms. |
| | $-$ | 8 | | | $-$ | 8 | Subtract 8 from both sides. |

| $2x$ | $=$ | x | $+$ | 3 | |
| $-x$ | | $-x$ | | | Subtract "x" from both sides. |

| x | $=$ | 3 |

Example 2: $x - 6 + 2x + 4 = 9 - 3x + 2x - 6$

Solution:

| $3x$ | $-$ | 2 | $=$ | 3 | $-$ | x | Combine like terms. |
| | $+$ | 2 | | $+2$ | | | Add 2 to both sides. |

| $3x$ | $=$ | 5 | $-$ | x | |
| $+x$ | | | $+$ | x | Add "x" to both sides. |

$\dfrac{4x}{4} = \dfrac{5}{4}$ Divide both sides by 4, the coefficient of "x".

$x = \dfrac{5}{4}$

Practice problems with variables on both sides of the equation. Solve the following.

1. $x - 4 + 3x + 1 = 4 - 4x + x + 1$

2. $-5x - 1 + 6x - 2 = 5 + 3x + 2x - 4$

3. $2x - 2 - 4x + 4 = -2 - 2x - 3x + 2$

4. $x - 2 + 6x + 4 = -2 + 4x + 2x - 1$

5. $3x - 1 - 5x + 2 = -4 - 4x - x - 5$

6. $5x + 4 - x - 2 = 4 - 4x + x + 5$

7. $3x - 2 - 6x + 3 = 3 + 2x - 3x - 4$

8. $3x - 1 - x - 1 = 5 - x - 4x + 1$

9. $-4x + 2 - 2x - 1 = 2 - 2x - 3x + 2$

10. $-3x - 5 - 5x + 1 = -2 + 2x - 4x - 1$

11. $-6x - 4 - 4x - 4 = -4 - 3x - x - 2$

12. $x - 2 - 2x - 5 = -3 + 2x + x - 5$

13. $-6x - 9 + 5x = 12x + 6 - 15x + 5$

14. $-5x + 12 + 4x - x = -3x - 6 - 5x$

15. $-x + 3x + 3x = -x - 20 + 2x$

16. $-3x - 12 - 5x + 2x = 2x + 4x$

17. $-4x + 8 + x - 5x = -4x - 2 + x$

18. $-x - 2 + 6x = -2x - 6 - 11x - 3$

19. $6x - 5 + 15x = -6x - 8 + 10x - 7$

20. $-6x + 39 - x + 7 = 94 + x$

21. $2x - 8 - 4x - 2 = 35 + 7x$

22. $6x + 10 - 5x - 10 = -20 - 4x$

23. $7x + 3x - 18 = -5x + 12$

24. $5x + 3x - 2x + 24 = 10x - 48$

25. $-6x + 12 + 8x = -5x + 54$

26. $3(2x) + 4 - 8 = (5 \cdot 4)$

27. $32 - 8x + 5x = -(6)^2 + 8x + 2$

28. $(3x + 2x + 5x) - (2x + x + 4x) = -(4^2 + 16 + 4)$

LESSON 11 – More Equations

Practice problems on more equations. Solve the following equations.

1. $-2x + 2 = -5x - 1$

2. $3x - 2 = x + 2$

3. $-5x - 3 = -4x + 4$

4. $-3x + 16 + 3x - 7 = 27 - 6x$

5. $x + 2 = 2x - 3$

6. $-4x - 1 = 5x - 5$

7. $-4x + 61 - 7x - 38 = 131 + 7x$

8. $2x - 2 = 3x - 5$

9. $5x - 4 = -4x + 5$

10. $-6x - 5 = x - 3$

11. $-3x - 4 = 5x - 5$

12. $x + 4 = -4x - 5$

13. $-6x - 5 = -x - 1$

14. $-3x + 6 - 4x - 9 = -27 - 3x$

15. $2x - 1 = 4x + 3$

16. $-2x + 1 = -5x + 5$

17. $5x + 2 = -2x + 1$

18. $2x + 8 + 3x + 33 = -13 - 4x$

19. $5x + 3 = 6x - 1$

20. $-3x + 4 = -4x + 2$

21. $-7x - 1 + x + 4 = -5 - 2x$

22. $4x - 4 - 6x + 3 = 4 - 7x$

23. $2x + 10 + 3x - 29 = 26 - 4x$

24. $6x + 16 + 4x - 4 = -12 + 6x$

25. $-5x + 12 + x = 5x - 2 - 3x + 10$

26. $2x + 5 = 8x - 25$

27. $-6x + 2x + 17 = 5x - 19$

28. $4x - 22 = -2x + 26$

LESSON 12 – Solve Decimal Equations

Today's topic is working with decimals in solving linear equation. We will work our way up the ladder in difficulty. So here it goes!

Example 1: $0.6x = 15$

    Solution: Clear out the decimal point by multiplying both sides of the equal sign by 10.

        $0.6x \cdot 10 = 15 \cdot 10$
        $6x = 150$ Divide both sides by 6.
        $x = 25$

    Check it out: $0.6(25) \overset{\checkmark}{=} 15$ AOK!

Example 2: $1.4x + 6 = -22$

    Solution: Again, multiply all three terms by 10.

        $1.4x(10) + 6(10) = (-22)(10)$
        $14x + 60 = -220$ Subtract 60 from both sides.
        $14x = -280$ Divide by sides by 14.
        $x = -20$

    Check it out: $1.4(-20) + 6 = -22$
                 $-28 + 6 = -22$
                 $-22 \overset{\checkmark}{=} -22$ AOK!

Note: Students sometime like to get variables and constants on opposite sides of the equation. It is your choice. See Example 3.

Example 3: $2.4x + 0.6x = 50.46 + 12.3$

    Solution: I suggest adding 2.4x and 0.6x on the left side. Also, adding 50.46 and 12.3 on the right side.

        $3x = 62.76$ Divide both sides by 3.
        $x = 20.92$

    Check it out: $2.4(20.92) + 0.6(20.92) = 50.46 + 12.3$
                 $50.208 + 12.552 = 62.76$
                 $62.76 \overset{\checkmark}{=} 62.76$ AOK!

Three for three! Keep on rolling!

Example 4: 3.02x + 15.4 = 1.6x + 24.275   (use your calculator)

Solution: Multiply **all** four terms by 1000 to remove all the decimals. Why 1000? Because 24.326 needs to move three places to the right. That would be 1000.

$$3.020\ x + 15.400 = 1.600\ x + 24.275$$

$$
\begin{array}{ccccc}
3020x & + & 15400 & = & 1600x & + & 24275 \\
 & - & 15400 & & & - & 15400 \\
\hline
\end{array}
$$
Subtract 15400 from both sides.

$$
\begin{array}{ccccc}
3020x & = & 1600x & + & 8875 \\
-1600x & & -1600x & & \\
\hline
\end{array}
$$
Subtract 1600x from both sides.

$$\frac{1420x}{1420} = \frac{8875}{1420}$$ Divide both sides by 1420.

$$x = 6.25$$

Check it out: 3.02(6.25) + 15.4 = 1.6(6.25) + 24.275
18.875 + 15.4 = 10 + 24.275
34.275 $\overset{\checkmark}{=}$ 34.275   AOK!

Example 5: 0.14x − 3.81 = 0.07x + 1.23

Solution:
Multiply all four terms by 100.

14x − 381 = 7x + 123

Subtract 7x from both sides and add 381 to both sides.

$$
\begin{array}{ccccc}
14x & - & 381 & = & 7x & + & 123 \\
- 7x & + & 381 & & -7x & + & 381 \\
\hline
\end{array}
$$

7x = 504   Divide both sides by 7.
x = 72

Check it out.

In this example, we have added and subtracted simultaneously.

33

Practice problems.

1. $2.1x - 3.2 = -8.4x - 45.2$

2. $6.4 - 4.2x = 16.8x + 90.4$

3. $2.418x = -12.09$

4. $34.3x - 5.95x = 283.5$

5. $2.18x + 6x - 4.3225 = 8.645 - 0.465x$

6. $30.7x - 18x = 76.2$

7. $39.4 = 7.32 + 3.208x$

8. $2.48x + 8x = 92.28 - 4.9x + 30.76$

9. $-4.1 = x + 1.1$

10. $x + (-1.3) = -1.5$

11. $-2x - 5.9 = 3.1 + x$

12. $2x - 0.53 = 1.42 - x$

13. $21.5 + 3.2x = 14.3 + 0.8x$

14. $-33.2 = -18.2 + 5x$

15. $16.4 - 7.1x = 28.46 - 1.07x$

LESSON 13 – Solving Fractional Equations

Another area of difficulty with algebra is solving equations containing fractions. **"FRACTIONS"** is the most dreaded word in mathematics.

Let us review several examples to alleviate those misapprehensions.

Now is the time to **_check_** your answer.

Example 1: $\frac{1}{3}x = 2$

> Solution: Multiply both sides of the equation by 3, the reciprocal of $\frac{1}{3}$.
>
> $3 \cdot \frac{1}{3}x = 2 \cdot 3$ The threes on the left cancel.
> $x = 6$
>
> Check your solution in the **original** equation: $\frac{1}{3} \cdot 6 \overset{\checkmark}{=} 2$

Example 2: $-\frac{2}{7}y = 10$

> Solution: Multiply both sides by the reciprocal of $-\frac{2}{7}$, which is $-\frac{7}{2}$.
>
> $\left(-\frac{7}{2}\right) \cdot \left(-\frac{2}{7}\right)y = 10 \cdot \left(-\frac{7}{2}\right)$ $\longrightarrow \left(-\frac{7}{2}\right) \cdot \left(-\frac{2}{7}\right) = 1$
>
> $y = \dfrac{10(-7)}{2} = -\dfrac{70}{2} = -35$
>
> Checking the solution: $-\frac{2}{7} \cdot (-35) = 10$
>
> $\dfrac{70}{7} = 10$
>
> $10 \overset{\checkmark}{=} 10$

Example 3: $x + \frac{1}{4} = 3$

Solution:

$x + \frac{1}{4} = 3$ Subtract $\frac{1}{4}$ from both sides.

$\underline{\quad -\frac{1}{4} \quad -\frac{1}{4} \quad}$

$x = 3 - \frac{1}{4} = 2\frac{3}{4}$

$3 = 2\frac{4}{4}$

$\underline{-\frac{1}{4} = -\frac{1}{4}}$

$2\frac{3}{4}$

Check: $2\frac{3}{4} + \frac{1}{4} = 3$

$3 \overset{\checkmark}{=} 3$

Example 4: $x - \frac{3}{5} = \frac{1}{10}$

Solution:

$x - \frac{3}{5} = \frac{1}{10}$ Add $\frac{3}{5}$ to both sides.

$\underline{\quad +\frac{3}{5} \quad +\frac{3}{5} \quad}$

reduce by 5

$x = \frac{1}{10} + \frac{3}{5} = \frac{(5)(1) + (3)(10)}{(10)(5)} = \frac{5 + 30}{50} = \frac{35}{50} = \frac{7}{10}$

$\frac{1}{10} = \frac{1}{10}$

$\underline{+\frac{3}{5} = +\frac{6}{10}}$

$\frac{7}{10}$

Check: $\frac{7}{10} - \frac{3}{5} = \frac{1}{10}$

$\frac{(7)(5) - (3)(10)}{(10)(5)} = \frac{1}{10}$

$\frac{35 - 30}{50} = \frac{1}{10}$

$\frac{5}{50} = \frac{1}{10}$

$\frac{1}{10} \overset{\checkmark}{=} \frac{1}{10}$

Example 5: $x + \frac{1}{5} = \frac{2}{3}$

Solution:

$$x + \frac{1}{5} = \frac{2}{3}$$
$$-\frac{1}{5} \quad -\frac{1}{5}$$

Subtract $\frac{1}{5}$ from both sides.

$$\frac{2}{3} = \frac{10}{15}$$
$$-\frac{1}{5} = -\frac{3}{15}$$
$$\frac{7}{15}$$

$$x = \frac{2}{3} - \frac{1}{5} = \frac{(2)(5) - (1)(3)}{(3)(5)} = \frac{10 - 3}{15} = \frac{7}{15}$$

Check it!! $\frac{7}{15} + \frac{1}{5} = \frac{2}{3}$

reduce by 25

$$\frac{(7)(5) + (1)(15)}{(15)(5)} = \frac{35 + 15}{75} = \frac{50}{75} = \frac{2}{3} \checkmark$$

Example 6: $\frac{1}{2}x + 3\frac{2}{3}x - 5 = 10 + 8 + 2$

Solution:

Step 1	$\frac{25}{6}x$	$- \quad 5 =$	20	Combine like terms.
Step 2		$+ \quad 5$	$+ 5$	Add 5 to both sides.

$$\frac{25}{6}x = 25$$

Step 3 $\frac{6}{25} \cdot \frac{25}{6}x = 25 \cdot \frac{6}{25}$ Multiply both sides by $\frac{6}{25}$, the reciprocal of $\frac{25}{6}$.

$$x = 6$$

Now try these to sharpen your math.

1. $\frac{2}{3}x = 14$

2. $\frac{4y}{3} = -2$

3. $-\frac{8x}{5} = -16$

4. $2\frac{1}{3}x = -14$ hint: change $2\frac{1}{3}$ to improper fraction

5. $-\frac{3}{5}x = 6$

6. $\frac{x}{7} = -5$

7. $x + \frac{2}{3} = 3\frac{1}{4}$

8. $x - 1\frac{1}{2} = 5\frac{2}{3}$

9. $6 - \frac{x}{3} = 4$

10. $5\frac{2}{3} - x = 16$

11. $a + \frac{3}{4} = 2$

12. $x - \frac{2}{5} = \frac{1}{10}$

13. $y + \frac{1}{3} = \frac{3}{5}$

14. $-2x = \frac{6}{7}$

15. $-\frac{5}{6} + y = \frac{7}{18}$

16. $x + \frac{2}{3} + \frac{5}{8} = 12$

17. $y - \frac{7}{10} + \frac{3}{5} = 4\frac{2}{5}$

18. $1\frac{1}{5}x + \frac{3}{10} - \frac{2}{5} = \frac{3}{5}x + 9\frac{2}{5} + \frac{7}{10} - \frac{x}{10}$

19. $-\frac{3}{4}x + \frac{2}{3} = \frac{1}{6}$

20. $1\frac{2}{3} + 1\frac{1}{2}x = -1\frac{1}{6}$

21. $1\frac{1}{3}x - 1\frac{1}{2} + \frac{1}{3}x = -1\frac{11}{12}$

22. $2\frac{5}{6}x + 1\frac{2}{3} + \frac{1}{3}x = -2\frac{1}{12}$

23. $-1\frac{3}{4} - 2\frac{1}{2}x = -3\frac{3}{8}$

24. $-\frac{1}{6}x + \frac{1}{12} = -\frac{3}{4}$

25. $\frac{1}{3}x + \frac{1}{2} = -\frac{3}{4}$

26. $-1\frac{1}{4}x - 3\frac{1}{6} + \frac{1}{6}x = -3\frac{1}{12}$

27. $3\frac{1}{3}x - 2\frac{1}{4} + \frac{1}{3}x = -1\frac{1}{12}$

28. $-2\frac{1}{6} - 1\frac{3}{4}x = -3\frac{7}{12}$

29. $\frac{1}{3}x - \frac{1}{4} = \frac{7}{12}$

30. $\frac{3}{5}x + 1\frac{2}{9} = 7\frac{4}{5}$

LESSON 14 – More Equations with Decimals and Fractions
One and Two Step Solutions

Example 1: $1.2x + 2.4x = 20.5 + 0.38$

Solution:

$3.6x = 20.88$ Combine like terms on both sides.

$$\frac{3.6x}{3.6} = \frac{20.88}{3.6}$$ Divide both sides by 3.6, the coefficient of "x".

$x = 5.8$

Example 2: $4x + \frac{1}{2} = 8\frac{3}{4} - \frac{1}{4}$

Solution:

$$4x \quad + \quad \tfrac{1}{2} \quad = \quad 8\tfrac{1}{2}$$ Combine like terms.

$$- \quad \tfrac{1}{2} \qquad - \tfrac{1}{2}$$ Subtract $\frac{1}{2}$ from both sides.

$$4x = 8$$

$$\frac{4x}{4} = \frac{8}{4}$$ Divide both sides by 4.

$$x = 2$$

Example 3: $3\frac{2}{3}y = 22 + 33$

Solution:

$$3\tfrac{2}{3}y = 55$$ Combine like terms.

$$\tfrac{11}{3}y = 55$$ Convert $3\frac{2}{3}$ to an improper fraction, $\frac{11}{3}$.

$$\tfrac{3}{11} \cdot \tfrac{11}{3}y = 55 \cdot \tfrac{3}{11}$$ Multiply both sides by $\frac{3}{11}$, the reciprocal of $\frac{11}{3}$.

$$y = 15$$

Example 4: $0.6x + 0.7x = 5 + \frac{1}{5}$

Solution:

$$1.3x = 5\tfrac{1}{5}$$ Combine like terms on both sides.

$$1.3x = 5.2$$ Convert $5\frac{1}{5}$ to a decimal.

$$\frac{1.3x}{1.3} = \frac{5.3}{1.3}$$ Divide both sides by 1.3, the coefficient of x.

$$x = 4$$

Practice on equations with decimals and fractions. Solve the following equations.

1. $-8h = 7.2$

2. $-4g = 3.2$

3. $-18k = -504$

4. $-15s = -345$

5. $\frac{4}{5}g - 7 = 1$

6. $\frac{2}{3}w - 5 = 7$

7. $3c + 14 = 29$

8. $7s - 18 = 3$

9. $4h + 39 = 47$

10. $0.5x - 1.6 = 1.2$

11. $0.5w - 1.2 = 1.2$

12. $1\frac{2}{3}h - 27 = 63$

13. $1\frac{4}{7}y - 22 = 55$

14. $2\frac{3}{5}p - 17 = 74$

15. $\frac{7}{9}y - 6 = 8$

16. $\frac{5}{6}y - 8 = 2$

17. $\frac{5}{6}y - 2 = 8$

18. $-9n = 6.3$

19. $4a + 38 = 70$

20. $9u - 32 = 13$

21. $-8w = 5.6$

22. $-9t = 5.4$

23. $4\frac{1}{3}b = 117\frac{2}{3}$

24. $0.2p + 1.2 = 1.5$

25. $2\frac{3}{4}N + 3.6 = 2\frac{2}{3}N + 19.85$

26. $5\frac{1}{2}x + 16 = 3\frac{3}{5}x - 60$

27. $1.76A + 3 - 0.60A = 10\left(\frac{1}{2}\right) + 0.32$

28. $\frac{4}{3}B + \frac{5}{8}B - 20.3 = 73.7$

40

LESSON 15 – Literal Equations

When solving a literal equation, letters replace most numbers. It can become tricky, so review the examples prior to tackling the practice problems.

Example 1:   Solve $\boxed{V = lwh}$ for "h".

> Solution:
> > Step 1:   Isolate the "h" to one side of the equation and move any coefficients to the opposite side. In V = lwh, the coefficients of "h" are "l" and "w". Therefore, divide ***both*** sides of the original equation by lw. You then have:
>
> $$\frac{V}{lw} = \frac{lwh}{lw}$$
>
> > Step 2:   Cancel the "lw" on the right side.
>
> > Step 3:   Final answer: $\dfrac{V}{lw} = h$

Example 2:   Solve $\boxed{A = \frac{1}{2}bh}$ for "b".

> Solution:
> > Step 1:   Divide both sides by the coefficient of "b". That would be $\frac{1}{2}h$.
>
> $$\frac{A}{\frac{1}{2}h} = \frac{\frac{1}{2}bh}{\frac{1}{2}h}$$
>
> > Step 2:   On the right side of the equation, the $\frac{1}{2}h$ in the numerator and the denominator cancel out.
>
> > Step 3:   Final answer: $\dfrac{A}{\frac{1}{2}h} = b$

Example 3:   Solve $\boxed{A = \dfrac{t-4}{r+s}}$ for "r".

Solution:

Step 1:   Multiply both sides by "r + s".

$$A(r + s) = t - 4$$
$$Ar + As = t - 4$$

Step 2:   Subtract "As" from both sides.

$$Ar = t - 4 - As$$

Step 3:   Divide both sides by "A".

$$\frac{Ar}{A} = \frac{t - 4 - As}{A}$$

Step 4:   Final answer: $r = \dfrac{t - 4 - As}{A}$

Example 4:   Solve $\boxed{A = p + prt}$ for "r".

Solution:

Step 1:   Subtract "p" from both sides of the equation.

$$A - p = (p - p) + prt$$
$$A - p = prt$$

Step 2:   Divide by the coefficient of "r", which is "pt".

$$\frac{A - p}{pt} = \frac{prt}{pt}$$

Step 3:   Remove the "pt" from the numerator and denominator on the right.

Step 4:   Final answer: $\dfrac{A - p}{pt} = r$

DO NOT TRY TO CANCEL THE "p's" ON THE LEFT!!
THAT'S A BIG NO-NO!

Now let us try these practice problems. Solve for the indicated variable.

1. Solve for "r": $E = ra$

2. Solve for "W": $A = LW$

3. Solve for "t": $i = prt$

4. Solve for "a": $S = \frac{1}{2}at^2$

5. Solve for "b": $S = a + b + c$

6. Solve for "W": $P = 2L + 2W$

7. Solve for "t": $A = p + prt$

8. Solve for "q": $\frac{D}{d} = q + \frac{r}{d}$

9. Solve for "A": $S = \frac{\pi r^2 A}{90}$

10. Solve for "g": $R = \frac{gs}{g + s}$

11. Solve for "f": $T = m(g - f)$

12. Solve for "R": $E = IR + Ir$

13. Solve for "q": $\frac{1}{f} = \frac{1}{p} + \frac{1}{q}$

14. Solve for "r": $\frac{D}{d} = q + \frac{r}{d}$

15. Solve for "k": $m = \frac{a - k}{6q}$

16. Solve for "r": $a = p(1 + rt)$

17. Solve for "R": $P - QR = TOP$

18. Solve for "W": $3H + \frac{2}{3}W = 5H - 9$

19. Solve for "x": $y(1 + x) = z - y$

20. Solve for "b": $abc = d(1 - ac)$

LESSON 16 – A Review of Lesson 1-15

1. Simplify: $18 - 3 \times 4 \div (6 - 4)$

2. Evaluate: $-3a + 4b$ if $a = -2$ and $b = 6$.

3. Solve: $2x + 5 = 27$

4. Evaluate: $-4x^2 + 2y^2$ if $x = 2$ and $y = 3$

5. If $x > 0$ and $y < 0$, which of the following is positive?
    a. xy b. $2xy$ c. $|x| \, y$ d. xy^2

6. Solve: $4(x + 1) = 28$

7. Solve for "h": $A = \frac{1}{2}bh$

8. Simplify: $2x - 3y + 4z - 7x + 5(y - 2z)$

9. Solve: $0.2x + 4.62 = -0.9x + 26.9225$

10. Simplify: $6[2a - 3b + |-8|] - \frac{5}{2}(4a - 8b)$

11. Of these expressions, which is undefined when x = 0?

   a. $\dfrac{x}{-3}$

   b. $\dfrac{x}{2x-3}$

   c. $\dfrac{-5}{x}$

   d. $\dfrac{x^3}{4}$

   e. $\dfrac{x+1}{x-1}$

   f. $\dfrac{5}{|x-4|}$

12. Solve: $2\frac{1}{2}x - 3\frac{5}{8} = 1\frac{3}{4}x + 5\frac{51}{80}$

13. Evaluate: $4(-m) + 3n - 2k$ if m = 2, n = 5, and k = x

14. Simplify: $[3(6-4) \div (2^2 \bullet 3)] + \dfrac{5}{2}$

15. True or False: $|-8| + |7| = |-7| + |8|$

16. Solve: $(7a - 2) - (4a - 5) = -2(a - 10)$

17. Simplify: $[-(-3)(-5)(4)] + [(-5)(-15)]$

18. Solve for "r": $C = 2\pi r$

19. Solve: $1\frac{3}{5}x + \frac{4}{9} = \frac{2}{3}x + 7\frac{2}{3}$

20. Solve: $5(2x - 3) = 85$

LESSON 17 – Circle and Line Graphs

Example 1:   The graph below shows sales in the million of dollars for XYZ Corporation. How much greater were sales in 2004 than 2001?

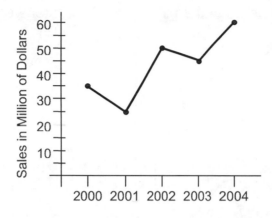

Solution:

2004 sales were $60 million. In 2001, sales were $25 million. Their difference ($60 million - $25 million = $35 million) is the answer.

Example 2:   The circle graph below shows the number of students who selected a beverage for lunch at H.O.W. School.

a. How many students selected soda for their beverage?
b. How many students did not select a beverage?
c. What percent of the students choose water as their beverage?

Solution:

a.   45

b.   10

c. $\dfrac{\text{water}}{\text{total number of students}} = \dfrac{25}{25 + 45 + 20 + 10} = \dfrac{25}{100} = 25\%$

46

Try these practice line and circle graph problems.

1. The graph in figure 17.1 below shows sales (in millions of dollars) for XYZ Corporation. How much greater were sales in 2004 than in 2001?
   a. $15,000,000
   b. $150,000
   c. $1,500,000
   d. $15

2. The graph in figure 17.2 below shows sales (in millions of dollars) for DEF Corporation. How much greater were sales in 2003 than in 2002?

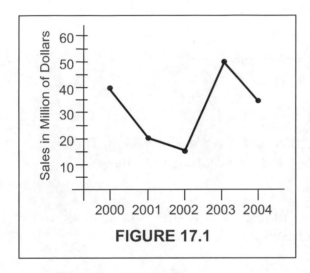

FIGURE 17.1

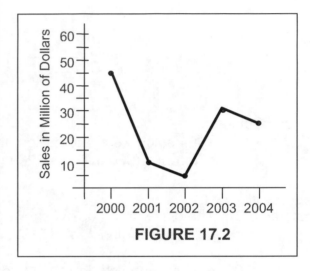

FIGURE 17.2

3. The graph in figure 17.3 below represents mortgage rates in April for several years. For what year was the interest rate the lowest?

4. The graph in figure 17.4 below shows sales (in millions of dollars) for HOC Corporation. How much greater were sales in 2000 than in 2003?

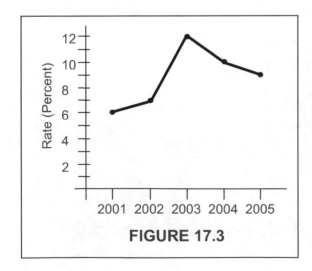

FIGURE 17.3

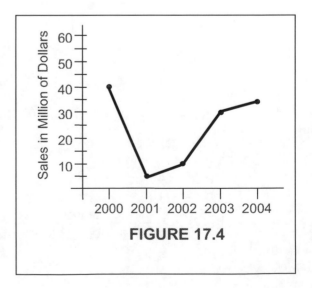

FIGURE 17.4

5. The graph in figure 17.5 below shows sales (in millions of dollars) for MOP Corporation.
   a. Which year had the most sales?
   b. Which year had the least sales?

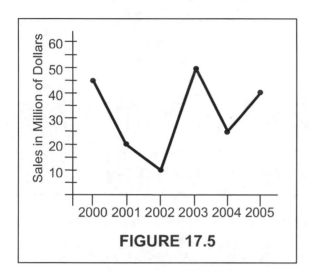

FIGURE 17.5

6. The graph in figure 17.6 below represents mortgage rates in March for several years.
   a. For what year was the interest rate the highest?
   b. What percent was that?

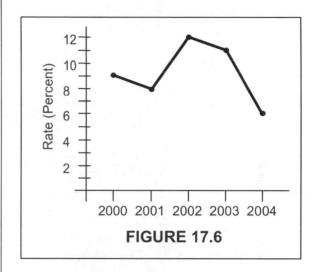

FIGURE 17.6

7. The circle graph in figure 17.7 below represents the number of students playing sports at Greyhound Junior High. How many more students play basketball than football?

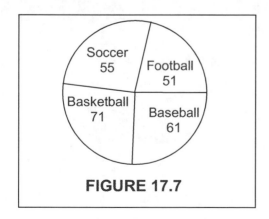

FIGURE 17.7

8. The circle graph in figure 17.8 below represents the number of students playing sports at Benson Elementary. How many students play soccer and football?

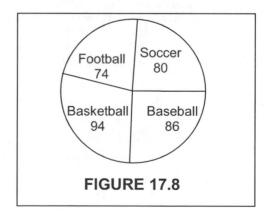

FIGURE 17.8

9. The circle graph in figure 17.9 below shows the number of students who chose each activity on Fun Day at Middle School. How many students chose the relay race?

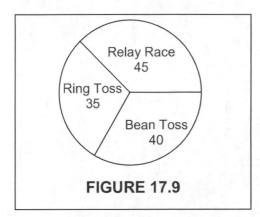

FIGURE 17.9

10. The circle graph in figure 17.10 below shows the number of students who chose each activity on Fun Day at Washington School. How many more students chose the bean toss over the ring toss?

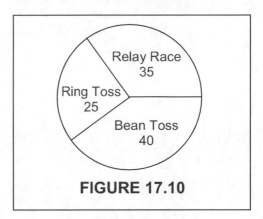

FIGURE 17.10

11. The circle graph in figure 17.11 below show the number of students playing sports at King Elementary School. How many more students play basketball than soccer?
a.　31　b.　143
c.　29　d.　121

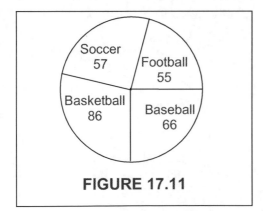

FIGURE 17.11

12. The circle graph in figure 17.12 below shows the number of students who chose each activity on Fun Day at Townsend School. How many students participated on Fun Day?

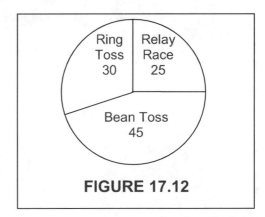

FIGURE 17.12

49

LESSON 18 – Linear Equations in Two Variables

A linear equation in two variables is an equation that can be written in the form:
$Ax + By = C$, where A, B, and C are real numbers and A and B are <u>not</u> both 0 (zero).

Some examples could be:

$$2x + 3y = 8 \qquad 4c - 5d = 32 \qquad a + b = 11 \qquad 3m - 2n = -17$$

Example 1:   Given the linear equation $3x + 2y = 7$, is the ordered pair (2, 3) a solution?

Solution: $3x + 2y = 7$

$3(2) + 2(3) \overset{?}{=} 7$

$6 + 6 \overset{?}{=} 7$

$12 \neq 7$

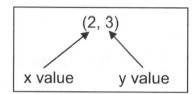

Therefore, (2, 3) is <u>not</u> a solution for $3x + 2y = 7$.

Try the ordered pairs for the given equation.

1. $4x + 2y = 18$ a. (2, 5)   b. (1, 8)   c. $(\frac{1}{2}, 8)$ d. (–5, 18)

2. $5x - 2y = 20$ a. (2, 5)   b. (3, –2)   c. (6, 5)   d. (10, 15)

3. $6x + y = 17$ a. (1, 9)   b. (2, 5)   c. (–3, 35)   d. $(\frac{1}{2}, 14)$

4. $3y - 2x = 8$ a. (2, 4)   b. (–3, 1)   c. (5, 10)   d. (4, 6)

Now we'll try a slight variation.

Example 2:   Given the equation $x + y = 9$, and the value of x is 3, what is the value of y?

Solution: $x + y = 9$

$3 + \underline{} = 9$

$3 + 6 = 9$

The ordered pair is (3, 6).

Example 3:   Complete the ordered pair (2, ___) given the equation 2x + y = 9.

Solution: 2x + y = 9
              2(2) + y = 9   replace x with 2
              4 + y = 9
              4 − 4 + y = 9 − 4   subtract 4 from both sides
              0 + y = 5
              y = 5

Check: 2(2) + 5 = 9
              4 + 5 = 9
              9 $\overset{\checkmark}{=}$ 9

The ordered pair is (2, 5).

Example 4:   Complete the ordered pair (3, ___) given the equation 5x − 2y = 13.

Solution: 5x − 2y = 13
              5(3) − 2y = 13   replace x with 3
              15 − 2y = 13
              15 − 15 − 2y = 13 − 15   subtract 15 from both sides
              0 − 2y = −2
              −2y = −2
              $\dfrac{-2y}{-2} = \dfrac{-2}{-2}$ divide both sides by −2
              y = 1

Check: 5(3) − 2(1) = 13
              15 − 2 = 13
              13 $\overset{\checkmark}{=}$ 13

The ordered pair is (3, 1)

Example 5:   Complete the ordered pair (___, −4) given the equation 8x − 3y = 36.

Solution: 8x − 3y = 36
              8x − 3(−4) = 36   replace y with −4 Note: −3(−4) = +12
              8x + 12 = 36
              8x + 12 − 12 = 36 − 12   subtract 12 from both sides
              8x + 0 = 24
              8x = 24
              $\dfrac{8x}{8} = \dfrac{24}{8}$ 8 divide both sides by 8
              x = 3

Check: $8(3) - 3(-4) = 36$
            $24 + 12 = 36$
            $36 \overset{\checkmark}{=} 36$

The ordered pair is (3, –4).

Let us try these. Complete the ordered pair to make the given equation true.

5.   (2, ___) $2x - y = 9$

6.   (1, ___) $5x - 2y = 13$

7.   (___, –4) $8x + 3y = 36$

8.   (2, ___) $-3x + 4y = 18$

9.   (___, 3) $y + 2x = 13$

10.   (6, ___) $5x + 2y = 22$

11.   (___, –8) $4x + 3y = -28$

Here's another variation to examine:

Example 6:   Given the equation $x + y = 7$, complete the table of order pairs.

	x	y
a.	3	
b.		5
c.	–1	
d.		–3

Solution: x + y = 7

    a.   x = 3, y = ?

        x + y = 7
        3 + y = 7   replace x with 3
        3 − 3 + y = 7 − 3   subtract 3 from both sides
        y = 4

        Check: 3 + 4 = 7
                   7 $\overset{\checkmark}{=}$ 7

    b.   x = ?, y = 5

        x + y = 7
        x + 5 = 7   replace y with 5
        x + 5 − 5 = 7 − 5   subtract 5 from both sides
        x = 2

        Check: 2 + 5 = 7
                   7 $\overset{\checkmark}{=}$ 7

    c.   x = −1, y = ?

        x + y = 7
        −1 + y = 7   replace x with −1
        −1 + 1 + y = 7 + 1   add 1 to both sides
        y = 8

        Check: −1 + 8 = 7
                   7 $\overset{\checkmark}{=}$ 7

    d.   x = ?, y = −3

        x + y = 7
        x + (−3) = 7   replace y with −3
        x − 3 = 7
        x − 3 + 3 = 7 + 3   add 3 to both sides
        x = 10

        Check: 10 + (−3) = 7
                   7 $\overset{\checkmark}{=}$ 7

To summarize, the table of ordered pairs is:

	x	y
a.	3	4
b.	2	5
c.	−1	8
d.	10	−3

Try these:

12.   Given the equation $2x - y = 9$, complete the table of ordered pairs.

	x	y
a.	1	
b.		3
c.		−5
d.	11	

13.   Complete the table of ordered pairs given the equation $3x + 2y = 16$.

	x	y
a.	2	
b.		2
c.	8	
d.		8

14.   Complete the table of ordered pairs given the equation $3y - 5x = 35$.

	x	y
a.	2	
b.	−4	
c.		20
d.		25

15.   Complete the table of ordered pairs given the equation 3x – y = –8.

	x	y
a.	1	
b.		–4
c.	4	
d.		–10

16.   Complete the table of ordered pairs given the equation 2x – 5y = 28.

	x	y
a.		–4
b.	–1	
c.	9	
d.		–8

16.   Complete the table of ordered pairs given the equation 4x + 5y = 17.

	x	y
a.	$\frac{1}{2}$	
b.		2
c.	3	
d.		4

16.   Complete the table of ordered pairs given the equation 3y – 2x = –12.

	x	y
a.		1
b.	2	
c.		0
d.	–3	

LESSON 19 – Cartesian Coordinate System

We now need to plot an ordered pair, (x, y), on the Cartesian Coordinate System.

Cartesian Coordinate System

II I
y
(−x, +y) | (+x, +y)
x
(−x, −y) | (+x, −y)
III IV

where

x is positive in the 1st and 4th quadrants.
y is positive in the 1st and 2nd quadrants.

Remember that the x-axis is horizontal and the y-axis is vertical.

Plot these ordered pairs (points) on a grid.

a.   (2, 3)
b.   (−3, −2)
c.   (4, 5)
d.   (−2, 3)
e.   (3, −2)
f.   (1, −3)
g.   (0, 4)

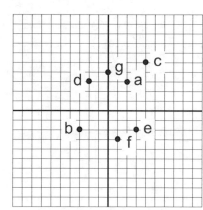

1.　Now it's your turn. Graph the following ordered pairs (points).

a.　(−3, 6)　　　e.　(6, −2)　　　h.　(1, 0)
b.　(2, 4)　　　f.　(−2, 4)　　　i.　(4, 1)
c.　(5, 0)　　　g.　(0, 3)　　　j.　(−4, 1)
d.　(−1, −2)

2. Let us try the reverse. State the ordered pair of the following points.

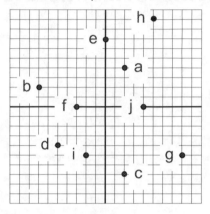

3. Either graph or state the ordered pair of points.

a.　(　,　)　　　e.　(　,　)　　　h.　(−4, 3)
b.　(5, 5)　　　f.　(−4, −3)　　　i.　(　,　)
c.　(2, −3)　　　g.　(　,　)　　　j.　(−1, −1)
d.　(　,　)

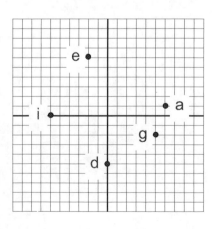

LESSON 20 – Graphing a Linear Equation

Put **A**x + **B**y = **C** into the form **B**y = –**A**x + **C** and solve for y:

$$y = -\frac{Ax}{B} + \frac{C}{B}$$

where the coefficient of x is the slope (m) and the constant is the y-intercept (b). This gives us the slope-intercept form of the equation:

$$y = mx + b$$

Example 1: Graph the equation 2x + 5y = 16.

Solution:
- Step 1: Solve for y.
- Step 2: Draw a grid.
- Step 3: Locate the y-intercept (b = $\frac{16}{5}$).

- Step 4: Play with the slope (m). "m" means $\frac{\Delta y}{\Delta x}$ or $\frac{\text{change in y}}{\text{change in x}}$. In this example, m = $-\frac{2}{5}$. For every –2 units on the y-axis there is a +5 units movement on the x-axis. Now you have located another point on the grid. Two points determine a line. Connect the points with a line and you have drawn the graph of that linear equation. Extend the line with rays from the two endpoints. Remember: a ray by definition is part of a line with one endpoint.

2x + 5y = 16

Subtract 2x from both sides.

5y = –2x + 16

Divide by 5.

$y = -\frac{2}{5}x + \frac{16}{5}$

$m = -\frac{2}{5}$ and $b = \frac{16}{5}$

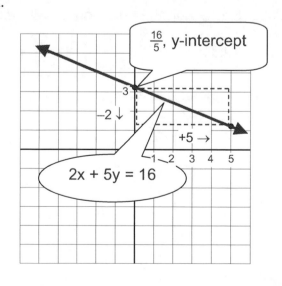

$\frac{16}{5}$, y-intercept

2x + 5y = 16

Example 2: Graph the equation 3x − 4y = 18.

Solution:
- Step 1: Solve for y.
- Step 2: Draw a grid.
- Step 3: Locate the y-intercept $(-\frac{9}{2})$.
- Step 4: Play with the slope $(\frac{3}{4})$. The change in "y" is +3 units up and the change in "x" is +4 units to the right.

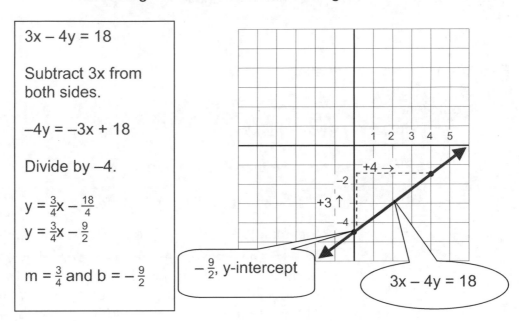

3x − 4y = 18

Subtract 3x from both sides.

−4y = −3x + 18

Divide by −4.

$y = \frac{3}{4}x - \frac{18}{4}$

$y = \frac{3}{4}x - \frac{9}{2}$

$m = \frac{3}{4}$ and $b = -\frac{9}{2}$

$-\frac{9}{2}$, y-intercept

3x − 4y = 18

Example 3: Graph the equation 3y − 2x = 9

Solution:

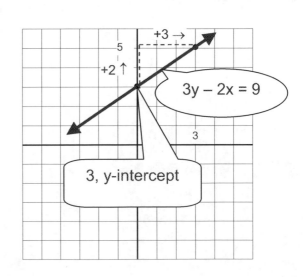

3y − 2x = 9

3y = 2x + 9

$y = \frac{2}{3}x + 3$

$m = \frac{2}{3}$ and $b = 3$

3y − 2x = 9

3, y-intercept

Example 4:   Graph the two equations, $\begin{cases} \mathbf{x - y = 3} \\ 2x + 3y = 8 \end{cases}$, on the **same** grid.

Solution:

x – y = 3
y = x – 3
m = 1 and b = –3
2x + 3y = 8
3y = –2x + 8
$y = -\frac{2}{3}x + \frac{8}{3}$
$m = -\frac{2}{3}$ and $b = \frac{8}{3}$

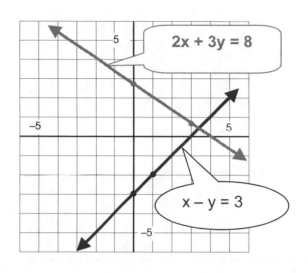

Example 5:   Graph the two equations, $\begin{cases} 2x - y = 5 \\ y - x = -3 \end{cases}$, on the **same** grid.

Solution:

2x – y = 5
y = 2x – 5
m = 2 and b = –5
y – x = –3
y = x – 3
m = 1 and b = –3

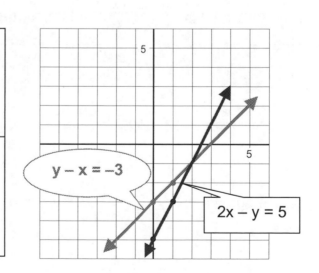

60

Example 6:   Two more equations to graph for good measure: $\begin{cases} 4x + y = 7 \\ -x + 2y = 8 \end{cases}$

Solution:

$4x + y = 7$
$y = -4x + 7$
$m = -4$ and $b = 7$
$-x + 2y = 8$
$2y = x + 8$
$y = \frac{1}{2}x + 4$
$m = \frac{1}{2}$ and $b = 4$

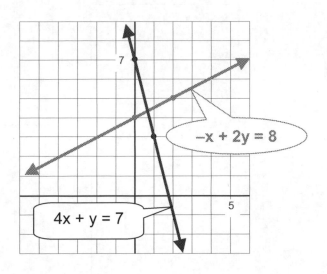

REMEMBER – Parallel lines have the same slope. Intersecting lines can be seen in example 5 above. Coincide lines are the same line.

Try some problems.

1.   Graph the equations $\begin{cases} y = -3x - 6 \\ y = 3x + 6 \end{cases}$ on the **same** grid. Are they parallel, intersect, or coincide?

2.   Graph the equations $\begin{cases} 2x - y = 6 \\ y - 2x = 6 \end{cases}$ on the same grid. Are they parallel, intersect, or coincide?

3.   Graph the equations $\begin{cases} y = 2x + 1 \\ y + 5 = 3x \end{cases}$ on the same grid. Are they parallel, intersect, or coincide?

4.   Graph the equations $\begin{cases} y = -\frac{3}{5}x + 2 \\ y = -\frac{2}{3}x \end{cases}$ on the same grid. Which equation goes through the origin?

LESSON 21 – Finding the Slope of a Line Given Two Points

The slope (m) of a line needs refinement.

$$m = \frac{\text{change in y}}{\text{change in x}} \text{ or } \frac{\Delta y}{\Delta x} = \frac{y_2 - y_1}{x_2 - x_1}$$

Let's look at this situation in a different way. Suppose you are given two points on a line, (2, 6) and (7, 15), what is the slope of the line?

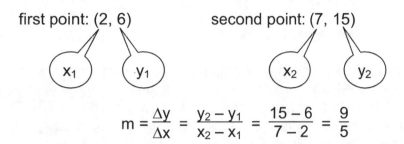

first point: (2, 6)   second point: (7, 15)

x_1 y_1 x_2 y_2

$$m = \frac{\Delta y}{\Delta x} = \frac{y_2 - y_1}{x_2 - x_1} = \frac{15 - 6}{7 - 2} = \frac{9}{5}$$

REMEMBER TO SELECT THE SAME POINT FOR ORDER PURPOSES!!

Practice. Try finding the slope of a line given two points.

1.   (3, 7) and (5, 8)   2.   (7, –3) and (6, 9)

3.   (–2, 6) and (8, 15)   4.   (–4, –6) and (–7, –9)

5.   (5, 5) and (8, 11)   6.   (4, 6) and (–2, 3)

7.   (2, 3) and (–1, –2)   8.   (–1, –5) and (–2, –7)

9.   (4, –3) and (–1, –8)   10.   (3, 1) and (5, 1)

11. $\left(-\frac{1}{4}, \frac{1}{4}\right)$ and $\left(\frac{1}{2}, -\frac{1}{2}\right)$ 12.   (2, 1) and (7, 3)

13.   (0, 0) and (1, –5)   14.   (3, –4) and (0, 0)

LESSON 22 – Find the Intercepts of a Given Line
Interpret the Slope and Graph the Line

Example 1:   Given the 2x + 3y = 12, find (a) the value of "x" and (b) the value of "y" respectively if one of the ordered pair is zero.

Solution:

(a)   When x = 0 and 0 is substituted into the equation 2x + 3y = 12, we get 2(0) + 3y = 12. Since 2 • 0 = 0, what remains is 3y = 12. By dividing both sides of 3y = 12 by 3, we arrive at y = 4. This is where the line intersects the y-axis when x = 0. So, when x = 0, y = 4 and the ordered pair is (0, 4).

(b)   When y = 0 and 0 is substituted into the equation 2x + 3y = 12, we get 2x + 3(0) = 12. Since 3 • 0 = 0, what remains is 2x = 12. By dividing both sides of 2x = 12 by 2, we arrive at x = 6. This is where the line intersects the x-axis when y = 0. So, when y = 0, x = 6 and the ordered pair is (6, 0).

Sometimes these points, (0, y) and (x, 0), are called the <u>zeros</u> of the equation.

Example 2:   Given the equation 2x – 3y = 18, find the zeros of the equation. Using these intercepts, graph the equation.

Solution:

a.   When x = 0, substitute 0 for "x": 2(0) – 3y = 18.
This gives us –3y = 18. Dividing both sides by –3, we arrive at y = –6. The y-intercept is (0, –6).

b.   When y = 0, substitute 0 for "y": 2x – 3(0) = 18.
This gives us 2x = 18. Dividing both sides by 2, we arrive at x = 9. The x-intercept is (9, 0).

Our two ordered pairs are (0, –6) and (9, 0). Locate the two points on a grid.

We can now graph the line of the equation 2x – 3y = 18 by connecting the two points.

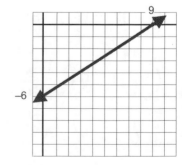

In what direction does the line go? It is either from "lower left to upper right" or "upper left to lower right". This will indicate the slope of the line. It is positive with reference to the former and negative referring to the latter.

+ (positive) if lower left to upper right
− (negative) if upper left to lower right

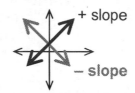

+ slope

− slope

You can also find the slope from the equation by putting in y = mx + b form. We will save that for Lesson 26, so let us practice what we have just examined.

Practice. Find the x- and y-intercepts (the zeros) of these equations.

1. $x + 3y = 9$

2. $2x - 3y = 10$

3. $-3x + 4y = 12$

4. $-2x - y = 6$

5. $4x + y = 10$

6. $-5x + 3y = 15$

7. $4x - 8y = -17$

8. $4x - 5y = 20$

9. $x + y = 2$

10. $x - y - 3 = 0$

11. $2x - y = -\frac{1}{2}$

12. $2x + 3y = 5$

13. $x - 4y + 3 = 0$

14. $5x + 1.5 = 3y$

LESSON 23 – Vertical and Horizontal Lines

What happens when the change in "x" (Δx) or "y" (Δy) is zero? Try these two points:

Example 1:   Find the slope given the points (6, –5) and (–12, –5).

Solution: $m = \dfrac{\Delta y}{\Delta x} = \dfrac{y_2 - y_1}{x_2 - x_1} = \dfrac{-5 - (-5)}{-12 - 6} = \dfrac{-5 + 5}{-18} = \dfrac{0}{-18} = 0$

Our slope is zero. We have a horizontal line that is parallel to the x-axis and the equation of this line is y = –5.

Now let us try these two points:

Example 2:   Find the slope given the points (–8, 6) and (–8, –1).

Solution: $m = \dfrac{\Delta y}{\Delta x} = \dfrac{y_2 - y_1}{x_2 - x_1} = \dfrac{-1 - 6}{-8 - (-8)} = \dfrac{-1 - 6}{-8 + 8} = \dfrac{-7}{0} =$ undefined!

Since we cannot divide by zero, our slope is undefined! We have a vertical line that is parallel to the y-axis and it's equation is x = –8.

In conclusion, when the "x" locations are **all** the same, the equation is x = ? with an undefined slope and **always** parallel to the y-axis. If the "y" values of both ordered pairs are identical, the equation is y = ? with a slope of zero and **always** parallel to the x-axis.

Try these problems! Find the equation of a line given the following information.

1.   (2, 4), (2, 8)

2.   (–8, 4), (–6, 4)

3.   (–9, –5), (–22, –5)

4.   (–3, 4), (–3, –8)

5.   (6, 0), (9, 0)

6. $\left(\frac{2}{3}, -\frac{5}{8}\right), \left(\frac{2}{3}, \frac{1}{8}\right)$

7.   (22, –10), (16, –10)

8.   (7, –3), (7, –8)

9. $\left(\frac{7}{9}, \frac{3}{4}\right), \left(16, \frac{3}{4}\right)$

10.   (–8, –8), (6, –8)

LESSON 24 – y = mx + b from Ax + By = C

The equation **A**x + **B**y = **C** can be rearranged to $y = -\dfrac{A}{B}x + \dfrac{C}{B}$. What can we conclude from this?

Example 1:   Solve 6x + 3y = 16 for y.

Solution: 6x + 3y = 16   subtract 6x from both sides
                 3y = –6x + 16   divide both sides by 3
                 $y = -2x + \dfrac{16}{3}$

                 our KEY indicators

Remember: y = mx + b
        where **"m"** (the coefficient of x) is the **slope** of the equation and **"b"** is the **y-intercept**. For the equation 6x + 3y = 16, the slope is –2 and the y-intercept is $\dfrac{16}{3}$.

With these two KEY pieces of information, we can graph, or draw, the line.

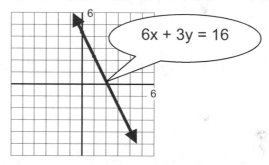

6x + 3y = 16

Grab some graph paper and try these. Find the zeros and the slope. Then graph the line.

1.   y = –3x

2.   2x – 3y = 18

3.   2y – 3x = 12

4.   x = 3

5. $\dfrac{4}{3}$x – 2y = 3
      hint: multiply all three terms by 3 to eliminate the fraction

6.   x – 2y = 2

7.   y = 5

8. $y = -\dfrac{2}{3}x + 5$

LESSON 25 – Determine Whether Lines are Parallel, Perpendicular or Neither

By definition, **parallel** lines have the **_same_** slopes, **Perpendicular** lines form **_right angles_**. Lines, in the same plane, that are not parallel nor perpendicular are intersecting (neither of the above).

Example 1: $\left.\begin{array}{l}\text{slope of line 1 is 2}\\ \text{slope of line 2 is 2}\end{array}\right\}$ lines are parallel

Example 2: $\left.\begin{array}{l}\text{slope of line 1 is }\frac{3}{5}\\ \text{slope of line 2 is }-\frac{5}{3}\end{array}\right\}$ lines are perpendicular

How do you find perpendicular lines? The product of the slopes of BOTH lines equals –1.

$$m_1 \cdot m_2 = -1$$

Example 3: $\left.\begin{array}{l}m_1 \text{ is }\frac{2}{3}\\ m_2 \text{ is }-\frac{3}{2}\end{array}\right\}$ $m_1 \cdot m_2 = \frac{2}{3} \cdot -\frac{3}{2} = -1$, lines are perpendicular

Now let's try real equations. Are these pairs of lines parallel, perpendicular or neither?

1. $2x + 5y = 4$
   $4x + 10y = 1$

2. $y = 2x - 3$
   $2y = -x + 4$

3. $-4x + 3y = 4$
   $-8x + 6y = 0$

4. $5x - 3y = -2$
   $3x - 5y = -8$

5. $8x - 9y = 6$
   $8x + 6y = -5$

6. $5x + 3y = 2$
   $3x - 5y = -1$

7. $3x - 2y = 6$
   $2x + 3y = 3$

8. $3x + 5y = 16$
   $2x - 4y = 7$

9. $y = 3x + 2$
   $y = x - 4$

10. $-\frac{1}{2}x - y = 0$
    $2x - y = 1$

**LESSON 26 – Finding the Equation of a Line Given Various Parameters
(Including the Point-Slope Form)**

A.   Slope and y-intercept: **y = mx + b**

    Example 1:   If the slope (m) of an equation is 2 and the y-intercept is 4, what is
                the equation of the line in standard form?

    Solution:
        Step 1:   In the equation y = mx + b, replace "m" with 2 and "b" with 4.

$$y = 2x + 4$$

        Step 2:   Put into standard form, **A**x + **B**y = **C**.

$$\textbf{A}x + \textbf{B}y = \textbf{C}$$

$$\left.\begin{array}{c} 2x - y = -4 \\ \text{or} \\ -2x + y = 4 \end{array}\right\} \text{ either is okay}$$

B.   Point-slope form: $y - y_1 = m(x - x_1)$ Very Important !!!!!

    Example 2:   If the slope (m) $= \frac{3}{4}$ and the point is (–4, –1), find the equation that
                satisfy these conditions.

    Solution:
$$y - y_1 = m(x - x_1)$$

$y - (-1) = \frac{3}{4}(x - (-4))$	replace "y_1" with –1, "m" with $\frac{3}{4}$, and x_1 with –4
$y + 1 = \frac{3}{4}(x + 4)$	remove negatives out of the equation
$4y + 4 = 3(x + 4)$	multiply all terms by 4, remember on the right side that the 4's cancel
$4y + 4 = 3x + 12$	expand the right side
$4y - 3x = 8$	collect like terms (constants)
$4y - 3x = 8$	
or	either equation is okay
$3x - 4y = -8$	

Try several. Find the equation of the line given the slope and a point on the line.
Answers should be in standard form and contain no fractions.

1.   (2, 7), m = 3   2.   (–4, 1), m = $\frac{3}{4}$

68

C.   Given two points: $m = \dfrac{\Delta y}{\Delta x} = \dfrac{y_2 - y_1}{x_2 - x_1}$ and $y - y_1 = m(x - x_1)$

This is a two-stepper. First find the slope, $m = \dfrac{\Delta y}{\Delta x}$, then use the point-slope formula, $y - y_1 = m(x - x_1)$, to find the equation in standard form with **_no_** fractions as coefficients.

Example 3:   Find the equation of a line given the points (4, 10) and (6, 12).

Solution:
   Step 1:   Find the slope.

$$m = \dfrac{\Delta y}{\Delta x} = \dfrac{y_2 - y_1}{x_2 - x_1} = \dfrac{12 - 10}{6 - 4} = \dfrac{2}{2} = 1$$

   Step 2:   Using m = 1 and **_either_** point, find the equation in standard form.

m = 1 and (4, 10)

$$y - 10 = 1(x - 4)$$
$$y - 10 = x - 4)$$
$$-x + y = 6$$

or

m = 1 and (6, 12)

$$y - 12 = 1(x - 6)$$
$$y - 12 = x - 6$$
$$-x + y = 6$$

As you can see, it doesn't matter which point you use, you will get the same equation!

Try these problems. Find the equation of the line given two points on the line. Answers should be in standard form and contain no fractions.

3.   (−4, 0) and (0, 2)   4. $\left(\frac{1}{2}, \frac{1}{3}\right)$ and $\left(-\frac{1}{4}, \frac{5}{4}\right)$ 5. $\left(-\frac{2}{3}, \frac{8}{3}\right)$ and $\left(\frac{1}{3}, \frac{7}{3}\right)$

LESSON 27 – Equation of a Line Satisfying Certain Parameters

Write an equation, in standard form, satisfying the given conditions.

Example 1: Passes through the point (2, –3) and parallel to the equation $3x - 4y = 5$.

Solution: Given a point and parallel to an equation, means we need to find the slope of the line, $\mathbf{m = \dfrac{\Delta y}{\Delta x}}$, and then use the point-slope formula $\mathbf{y - y_1 = m(x - x_1)}$.

$$3x - 4y = 5$$
$$-4y = -3x + 5$$
$$y = \tfrac{3}{4}x - \tfrac{5}{4}$$

$$m = \tfrac{3}{4}$$

Substitute $m = \tfrac{3}{4}$ and the point (2, –3) into the point-slope formula.

$$y - (-3) = \tfrac{3}{4}(x - 2)$$
$$y + 3 = \tfrac{3}{4}(x - 2)$$
$$4y + 12 = 3(x - 2)$$
$$4y + 12 = 3x - 6$$
$$18 = 3x - 4y$$

in standard form: $\boxed{3x - 4y = 18}$

Example 2: Passes through the point (–1, 4) and perpendicular to $2x + 3y = 8$.

Solution: First find the slope of the equation.

$$2x + 3y = 8$$
$$3y = -2x + 8$$
$$y = -\tfrac{2}{3}x + \tfrac{8}{3}$$

Remember for perpendicular lines, that: $m_1 \cdot m_2 = -1$.

$$-\tfrac{2}{3} \cdot m_2 = -1$$
$$-\tfrac{3}{2} \cdot -\tfrac{2}{3} \cdot m_2 = -1 \cdot -\tfrac{3}{2}$$
$$m_2 = \tfrac{3}{2}$$

Given $m = \frac{3}{2}$ and the point $(-1, 4)$, substitute into the point-slope formula.

$$y - y_1 = m(x - x_1)$$
$$y - 4 = \frac{3}{2}(x - (-1))$$
$$y - 4 = \frac{3}{2}(x + 1)$$
$$2y - 8 = 3(x + 1) \quad \text{multiply by 2}$$
$$2y - 8 = 3x + 3 \quad \text{distribute the 3}$$
$$-11 = 3x - 2y$$

in standard form: $\boxed{3x - 2y = -11}$

Example 3: Perpendicular to $x - 2y = 7$ with y-intercept $(0, -3)$.

Solution: First find the slope of the equation.

$$x - 2y = 7$$
$$-2y = -x + 7$$
$$y = \frac{1}{2}x - \frac{7}{2}$$

Remember for perpendicular lines, that: $m_1 \cdot m_2 = -1$.

$$\frac{1}{2} \cdot m_2 = -1$$
$$\frac{2}{1} \cdot \frac{1}{2}m_2 = -1 \cdot \frac{2}{1}$$
$$m_2 = -2$$

Given $m = -2$ and the point $(0, -3)$, substitute into the point-slope formula.

$$y - y_1 = m(x - x_1)$$
$$y - (-3) = -2(x - 0)$$
$$y + 3 = -x \quad \text{distribute the } -2$$
$$3 = -x - y$$

in standard form: $\boxed{2x + y = -3}$

Example 4: Parallel to 5x = 2y + 10 with y-intercept (0, 4).

Solution: Find the slope of the equation.

$$5x = 2y + 10$$
$$5x - 10 = 2y$$
$$\tfrac{5}{2}x - 5 = y$$

Remember that the slopes of parallel lines are the same. Therefore the slope of the desired line is m = $\frac{5}{2}$.

Using m = $\frac{5}{2}$ and the point (0, 4), substitute into the point-slope formula.

$$y - y_1 = m(x - x_1)$$
$$y - 4 = \tfrac{5}{2}(x - 0)$$
$$2y - 8 = 5(x - 0) \quad \text{multiply by 2}$$
$$2y - 8 = 5x \quad \text{distribute the 5}$$
$$-8 = 5x - 2y$$

in standard form: $\boxed{5x - 2y = -8}$

Let us try some problems. Write the equation in standard form with the given characteristics.

1. with a slope of 2, passing through the point (1, 4)

2. parallel to the x-axis and 8 units below it

3. parallel to 2x + y = 7 passing through the origin

4. perpendicular to x – 2y = 4 passing through (5, 8)

5. with integral coefficients whose slope is $-\frac{5}{6}$ and y-intercept is $-\frac{2}{3}$

6. parallel to 4x = –5y + 6 passing through (–2, 6)

7. perpendicular to 2x – 5y = 12 passing through (1, –3)

LESSON 28 – Graphing Linear Inequalities in Two Variables

This is just like graphing linear equalities except you need to shade in certain areas of the grid.

Example 1: Graph the inequality $x + 2y \geq 6$.

Solution: Temporary substitute the inequality sign ($\geq$) with an equal sign (=) and solve for y.

$$x + 2y = 6$$
$$2y = -x + 6$$
$$y = -\tfrac{1}{2}x + 3$$

Graph the equality. If the inequality is $\leq$ or $\geq$, keep a solid line. If the inequality is < or >, use a dashed line.

To decide which side of the line to shade, select the origin, (0, 0), and substitute in the **original inequality**.

$$0 + 2(0) \overset{?}{\geq} 6$$
$$0 + 0 \overset{?}{\geq} 6$$
$$0 \ngeq 6$$

Therefore, (0, 0) is not part of the graph and you need to shade the other half-plane.

It is a simple procedure:

Step 1: Graph as an equality.
Step 2: Solid or dashed line.
Step 3: Use (0,0) to find out if (0, 0) is in the shaded region
Step 4: Shade the appropriate half-plane.

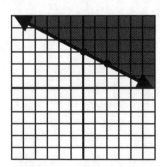

Example 2:   y < –3x + 1

Solution: y < –3x + 1 as an equality is in the slope-intercept form, y = mx + b.
               Therefore, m = –3 (slope) and b = 1 (y-intercept).

    Step 1: Graph as an equality.
    Step 2: Dashed line (<).
    Step 3: (0, 0) is in the shaded region.
    Step 4: Shade the appropriate region.
    Step 5: Check by selecting any point in the shaded region.

m = –3		Check:

m = –3
b = 1 (y-intercept)

plot "b" on the y-axis

slope = $\frac{\Delta y}{\Delta x}$ = $\frac{-3}{1}$

down 3, right 1

dashed line

substitute (0, 0)
     y < –3x + 1
     0 < 0 + 1
        0 < 1

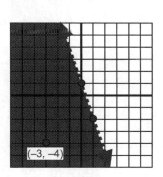

Check:

Select any point in
the shaded region.

(–3, –4)

Substitute into
original inequality.

y = < –3x + 1
–4 < –3(–3) + 1
   –4 < 9 + 1
      –4 < 10

1.   y ≤ 2x + 3

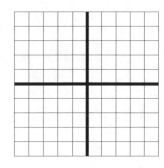

2.   y > –2x + 4

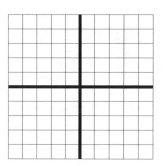

3.   y < –3x – 1

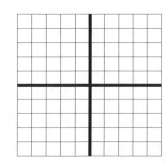

4.   y ≥ x – 5

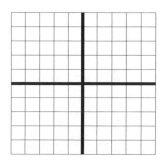

LESSON 29 – A Review of Lessons 17-28

1. Is the line $y = -2x + 3$ parallel to the line $4x - 2y = 5$?

2. Write the equation of a straight line, with integral coefficients, whose slope is $\frac{3}{4}$ and y-intercept is 5.

3. Does the graph of the equation $3x = 4y + 4$ go through the point (2, 5)?

4. What is the x-intercept of the graph of the equation $y = x - 6$?

5. If the graph of $y = mx + 4$ passes through the point (2, 7), find the value of "m".

6. Find two values of "y" corresponding to the "x" that could be used in plotting the graph of $x + 3y = 5$.

x	−1	5
y		

7. If the line, $y = 3x + 5$, is parallel to $y - 3x = 1$, find the similar slope.

8. Find the intercepts of the linear equation: $2x + 5y = 8$

9. Write the equation of the line parallel to $y = 3x - 1$ passing through the point (2, 4).

10. Is the equation, $y = 5$, a vertical or horizontal line?

11. Is the point (−3, −4) in the graph of the inequality, y = 2x + 4?

12. Write the equation, in standard form, perpendicular to 2x − 5y = 8 passing through the point (−2, 5).

13. From the information in question 12, does this line go through the origin?

14. Given the slope is 3 and going through the point (7, −1), find the linear equation in standard form.

15. Are these three points, A(−1, −4), B(−5, 8) and C(−2, −7), in the same quadrant?

16. Find the slope of the line through points A(−1, 4) and B(2, −5).

17. Find the x- and y-intercepts of this line: 3x = 14 − 5y

18. True or False: If x = 4, one possible solution of a linear equation is (4, 8).

19. True or False: The graph of the equations 2x + 4y = 8 and 5x − 5y = 5 **both** go through the origin.

20. (a) Given the circle graph to the left, find x%.
    (b) If 120 people are represented in the circle graph, how many are in the A part?

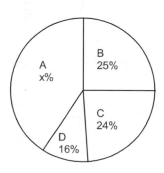

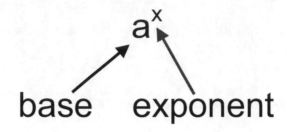

A.   [Product Rule]   When the bases are the same, add the exponents.

$$a^x \cdot a^y = a^{x+y}$$

Example 1: $4^3 \cdot 4^2 = 4^{3+2} = 4^5$

B.   [Power Rule]   Multiply the exponents.

$$(a^x)^y = a^{x \cdot y} = a^{xy}$$

Example 2: $(4^2)^3 = 4^{2 \cdot 3} = 4^6$ Example 3: $(c^2)^4 = c^{2 \cdot 4} = c^8$

C.   [Quotient Rule]   When the bases are the same, subtract the exponents.

$$\frac{a^x}{a^y} = a^{x-y}$$

Example 3: $\dfrac{m^5}{m^2} = m^{5-2} = m^3$

D.   [Zero Rule]   Anything, other than zero, to the zero power is 1.

Example 5: $x^0 = 1$ Example 7: $-4^0 = -1$ $(-1 \cdot 4^0 = -1 \cdot 1 = -1)$

Example 6: $5^0 = 1$ Example 8: $(-4)^0 = 1$

E. | Negative Exponent Rule | $a^{-x} = \dfrac{1}{a^x}$

Example 9: $4^{-2} = \dfrac{1}{4^2} = \dfrac{1}{16}$

Example 12: $4x^{-3} = \dfrac{4}{x^3}$

Example 10: $(x^2)^{-2} = \dfrac{1}{(x^2)^2} = \dfrac{1}{x^{2 \cdot 2}} = \dfrac{1}{x^4}$

or $\quad (x^2)^{-2} = x^{2 \cdot -2} = x^{-4} = \dfrac{1}{x^4}$

Example 13: $2xy(5x^{-2}y + 6a^{-2}x^2y) =$

$$\dfrac{10xy^2}{x^2} + \dfrac{12x^3y^2}{a^2}$$

$$\dfrac{10y^2}{x} + \dfrac{12x^3y^2}{a^2}$$

Example 11: $8^{-3} = \dfrac{1}{8^3} = \dfrac{1}{512}$

Evaluate $x^{-1}y^2$ when x = 2 and y = 3.

Solution: $\quad x^{-1}y^2 = \dfrac{y^2}{x} = \dfrac{3^2}{2} = \dfrac{9}{2}$ or $x^{-1}y^2 = 2^{-1} \cdot 3^2 = \dfrac{1}{2} \cdot 9 = \dfrac{9}{2}$

Rewrite $\dfrac{x^2 x^3 y^{-2} z^4}{x^{-2} y^3 z}$ with all positive exponents.

Solution: $\quad \dfrac{x^2 x^3 y^{-2} z^4}{x^{-2} y^3 z} = \dfrac{x^2 x^3 x^2 z^4}{y^3 y^2 z} = \dfrac{x^{2+3+2} z^4}{y^{3+2} z} = \dfrac{x^7 z^4}{y^5 z} = \dfrac{x^7 z^{4-1}}{y^5} = \dfrac{x^7 z^3}{y^5}$

Rewrite $\dfrac{x^2 x^{-4} y^2 z^{-3}}{x^3 y^{-3} z^5}$ with all variables in the numerator.

Solution: $\quad \dfrac{x^2 x^{-4} y^2 z^{-3}}{x^3 y^{-3} z^5}$

$(x^2 x^{-4} x^{-3}) \cdot (y^2 y^3) \cdot (z^{-3} z^{-5}) =$
$(x^{2-4-3}) \cdot (y^{2+3}) \cdot (z^{-3-5}) =$
$x^{-5} y^5 z^{-8}$

Distributive Property distributes multiplication over addition. Some examples are

$$a(b + c) = a \cdot b + a \cdot c = ab + ac$$
$$6(3 + x) = 6 \cdot 3 + 6 \cdot x = 18 + 6x$$
$$-3(x + 4y) = -3 \cdot x + (-3) \cdot 4y = -3x + (-12y) = -3x - 12y$$
$$5x(x + 3) = 5x \cdot x + 5x \cdot 3 = 5x^2 + 15x$$
$$7(a + 2b + c) = 7 \cdot a + 7 \cdot 2b + 7 \cdot c = 7a + 14b + 7c$$

Practice.

1. Expand $6g^4h^3(5g^{-2}h^2 - 2g^2h^{-2})$ by using the distributive property. Write answer with positive exponents.

2. Expand $4b^5c^3(2b^{-1}c^2 + 3b^2c^{-1})$ by using the distributive property. Write answer with positive exponents.

3. Multiply: $a^{-4}(a^2)(a^{-5})$ Select your answer.

   a. a^7 b. a^{40} c. $\dfrac{1}{a^8}$ d. $\dfrac{1}{a^7}$

4. Simplify.

   a. $\dfrac{1}{(-2)^{-2}}$ b. $(-2)^{-2}$ c. $(-2)^0$

5. Evaluate $x^{-2}y^0$ when $x = 5$ and $y = 4$.

   a. $\dfrac{1}{25}$ b. 0   c. -10 d. $\dfrac{1}{64}$

6. Simplify and write answer with all positive exponents.

   a. $\dfrac{x^3 \, x^5 \, y^3 \, z^{-3}}{z^{-2} \, z^{-2} \, y^{-3}}$ b. $\dfrac{x^5 \, x^3 \, y^5 \, z^{-3}}{z^{-1} \, z^{-1} \, y^{-5}}$

7. Simplify and write answer with all variables in the numerator.

   a. $\dfrac{x^{-4} \, x^3 \, y^2 \, z^{-4}}{z^{-1} \, z^{-2} \, y^{-4}}$ b. $\dfrac{x^{-5} \, x^4 \, y^3 \, z^{-2}}{z^{-2} \, z^{-1} \, y^{-1}}$ c. $\dfrac{33x^{-12} \, y^2 \, z^{20}}{11x^{-3} \, y^8 \, z^{-5}}$

Use the Distributive Property to simplify.

8. $a(x + 2y)$

9. $7(2 + 3x)$

10. $-3y(y^2 + 3z)$

11. $2x^2(x^2 + 3x + 4)$

12. $-5c(d^2 + ef)$

13. $4x(2x + 3)$

14. $-10(2a + 4b)$

15. $8a(5a^2 + 6a + 4)$

16. $-(m + n + 2p)$

17. $15(2 + 6x)$

18. $6m(m^2 + 3m + 6)$

19. $-9e(ef + eg)$

20. $-7x(x + y)$

21. $4x(3x^2 + 2x + 4)$

22. $2(3x + 4y + 5z)$

23. $-5t(t^3 + 3t + 4)$

24. $4xy(3y + 7xz)$

25. $3a(4b^{-2} - 7a^2c^{-2})$

26. $3x(4x^2 - 5x + 6)$

27. $8(\frac{9}{2}x^2 - \frac{3}{4}x + \frac{1}{8})$

28. $-x^2(x^3 - 4x^2 - 3x + 7)$

29. $24(\frac{x^2}{2} + \frac{x}{8} - \frac{5}{6})$

30. $2x[3x^2 + 6(2x + 3)]$

31. $10y(10y^2 - 6y - \frac{3}{2y})$

32. $-a(a^2 - ab - 5b^2)$

33. $-8e(4e^3 - 2e^2 + 3 + e)$

34. $ab(c^2 - 3c + 4)$

35. $9f(-f^2 + 4f - \frac{1}{3})$

LESSON 31 – Addition and Subtraction of Polynomials

There are two methods for this operation. Either perform the operation horizontally or vertically. Let me show you both ways.

HORIZONTALLY

Example 1: $(4a^2 + 3a - 6) + (3a + 8 - 2a^2)$

Solution: $4a^2 + 3a - 6 + 3a + 8 - 2a^2 =$ Remove grouping symbols.
$4a^2 - 2a^2 + 3a + 3a - 6 + 8 =$ Group like terms.
$(4a^2 - 2a^2) + (3a + 3a) + (-6 + 8)$ Combine like terms.
$2a^2 + 6a + 2$

Example 2: $3(x2 - 3x - 4) + (x - 5) - 2(x2 + 5x)$

Solution: $3x^2 - 9x - 12 + x - 5 - 2x^2 - 10x =$ Use Distributive Property.
$3x^2 - 2x^2 - 9x + x - 10x - 12 - 5 =$ Group like terms.
$(3x^2 - 2x^2) + (-9x + x - 10x) + (-12 - 5) =$ Combine like terms.
$x^2 - 18x - 17$

VERTICALLY – only after you use the Distributive Property.

Example 3: $(3x^3 + 16x^2 - 5x - 10) + (x^3 - 9x^2 + 6x - 4)$

Solution:

$$
\begin{array}{rrrrrrr}
 & 3x^3 & + & 16x^2 & - & 5x & - & 10 \\
+ & x^3 & - & 9x^2 & + & 6x & - & 4 \\
\hline
 & 4x^3 & + & 7x^2 & + & x & - & 14
\end{array}
$$

Example 4: $(16x^3 - 14x^2 - 7x + 8) + (-8x^3 - 2x^2 + 9x - 6)$

Solution:

$$
\begin{array}{rrrrrrr}
 & 16x^3 & - & 14x^2 & - & 7x & + & 8 \\
+ & -8x^3 & - & 2x^2 & + & 9x & - & 6 \\
\hline
 & 8x^3 & - & 16x^2 & + & 2x & + & 2
\end{array}
$$

Example 5: $(4a^3 - 4a^2 + 6a - 7) - (2a^3 - 6a^2 + 5a + 7)$

Solution: Remember in subtraction the signs in the subtrahend are changed.

$$4a^3 - 4a^2 + 6a - 7$$
$$+\ -2a^3 + 6a^2 - 5a - 7$$
$$2a^3 + 2a^2 + a - 14$$

Example 6: $(3x^2 + 4x - 11) + (2x^2 - 4x + 10) - (3x^2 - 5x + 6)$

Solution:

Vertically:

$$3x^2 + 4x - 11 \quad \text{To start, add the first two trinomials.}$$
$$+\ 2x^2 - 4x + 10$$
$$5x^2 + 0 - 1$$
$$+\ -3x^2 + 5x - 6 \quad \text{Then subtract the third trinomial.}$$
$$2x^2 + 5x - 7$$

Horizontally:

$3x^2 + 4x - 11 + 2x^2 - 4x + 10 =$ Remove grouping symbols from first
$3x^2 + 2x^2 + 4x - 4x - 11 + 10 =$ two trinomials, group and combine
$5x^2 - 1$ like terms.

$5x^2 - 1 - 3x^2 + 5x - 6 =$ Now subtract the third trinomial,
$5x^2 - 3x^2 + 5x - 1 - 6 =$ remembering to change the signs.
$2x^2 + 5x - 7$ Group and combine like terms.

or, remove the grouping symbols from all three trinomials, group and combine like terms:

$3x^2 + 4x - 11 + 2x^2 - 4x + 10 - 3x^2 + 5x - 6 =$
$(3x^2 + 2x^2 - 3x^2) + (4x - 4x + 5x) + (-11 + 10 - 6) =$
$2x^2 + 5x + (-7) =$
$2x^2 + 5x - 7$

PRACTICE MAKES PERFECT!

Add. Try using either the horizontal or vertical method. Write answer in descending order of the variable.

1. $(5x + 7x^3 - 3) + (7x^3 - 9 + 7x)$

2. $(3x^3 + 3x^2 - 3x - 3) - (5x^3 - 9x^2 - 7)$

3. $(7a^5 - 5a^2 - 9a + 5) - (3a^5 + 7a^2 + 3)$

4. $(7y^4 + 5y^2 + 9) - 3(5y^4 - 3y^2 + y - 7)$

5. $(-3x^5 + 7x^3 + 5) + 5(-9x^5 - 9x^3 + x - 5)$

6. $2(3x^5 + 2x^4 + 5x^3 + 6x^2 - x - 4) + 3(3x^4 + 2x^3 + 5x^2 + 6x - 1)$

7. $4(-4x^5 + 3x^4 - 2x^3 + 5x^2 + 6x - 1) - 2(4x^4 + 3x^3 - 2x^2 - 5x + 6)$

8. $(-5x^2 - x + 5) + (-4x^2 - 5x - 2)$

9. $(-z^2 - 2z + 2) + (5z^2 + 2z - 3)$

10. Subtract:
$$\begin{array}{r} 12a^4 - 15a^3 + 16a^2 + 8a - 14 \\ -\ 3a^4 +\ \ 5a^3 -\ \ 2a^2 + 6a - 11 \\ \hline \end{array}$$

11. $(6x^2 - 5x - 8) - (4x^2 + 7x - 2)$

12. $3(3n^2 - 6n + 4) + 2(-n^2 + 7n - 8)$

13. $(4a^3 - 6a^2 + 8a - 1) - (2a^2 - 6a + 9)$

14. $2y^3 + 5y^2 - 9y - 13 + 7y^2 - 8y^3 + 14$

LESSON 32 – Multiplication of Binomials – FOIL

First   Outer   Inner   Last

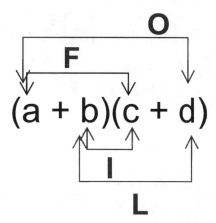

First terms: $a \cdot c = ac$
Outer terms: $a \cdot d = ad$
Inner terms: $b \cdot c = bc$
Last terms: $b \cdot d = bd$

$(a + b)(c + d) = ac + ad + bc + bd$

Example 1: $(2a + 3b)(c - 4d)$

   Solution: First terms: $2a \cdot c = 2ac$
                     Outer terms: $2a \cdot (-4d) = -8ad$
                     Inner terms: $3b \cdot c = 3bc$
                     Last terms: $3b \cdot (-4d) = -12bd$

                     $(2a + 3b)(c - 4d) =$
                     $2ac - 8ad + 3bc - 12bd$

Example 2: $(-4x + 3y)(x - 5y)$

   Solution: First terms: $(-4x) \cdot x = -4x^2$
                     Outer terms: $(-4x) \cdot (-5y) = 20xy$
                     Inner terms: $3y \cdot x = 3xy$
                     Last terms: $3y \cdot (-5y) = -15y^2$

                     $(-4x + 3y)(x - 5y) =$
                     $-4x^2 \textbf{ + 20xy + 3xy} - 15y^2 =$

                     combine like terms
                     $-4x^2 + 23xy - 15y^2$

Example 3: $(6x - 5y)(-2x + 3y)$

Solution: First terms: $6x \cdot (-2x) = -12x^2$
Outer terms: $6x \cdot 3y = 18xy$
Inner terms: $(-5y) \cdot (-2x) = 10xy$
Last terms: $(-5y) \cdot 3y = -15y^2$

$(6x - 5y)(-2x + 3y) =$
$-12x^2 \textbf{ + 18xy + 10xy} - 15y^2 =$
$-12x^2 + 28xy - 15y^2$

Example 4: $(4x + 3)^2$

Solution: $(4x + 3)^2 = (4x + 3)(4x + 3)$

F: $4x \cdot 4x = 16x^2$
O: $4x \cdot 3 = 12x$
I: $3 \cdot 4x = 12x$
L: $3 \cdot 3 = 9$

$(4x + 3)^2 =$
$16x^2 + 12x + 12x + 9 =$
$16x^2 + 24x + 9$

Example 5: $(2x - 5)(6x + 1)$

Solution: F: $2x \cdot 6x = 12x^2$
O: $2x \cdot 1 = 2x$
I: $-5 \cdot 6x = -30x$
L: $-5 \cdot 1 = -5$

$(2x - 5)(6x + 1)$
$12x^2 + 2x - 30x - 5 =$
$12x^2 - 28x - 5$

Example 6: $(a - 2b)(-3a + 5b)$

Solution: F: $a \cdot (-3a) = -3a^2$
O: $a \cdot 5b = 5ab$
I: $(-2b) \cdot (-3a) = 6ab$
L: $(-2b) \cdot 5b = -10b2$

$(a - 2b)(-3a + 5b) =$
$-3a^2 + 5ab + 6ab - 10b^2 =$
$-3a^2 + 11ab - 10b^2$

Now practice multiplying these binomials.

1. $(4m + n)(-m - 2n)$

2. $(x^2 + x)(3x^2 - 2x)$

3. $(7x + 3)(4x - 5)$

4. $(6a + 7b)(-2a + 3b)$

5. $(8c - 3d)(2c + 3d)$

6. $(3y + z)^2$

7. $(c - 5d)(-2c + d)$

8. $(-3a + 4)(7a - 2)$

9. $(4e - 3f)(6e + f)$

10. $(-8x - 3y)(3x + 4y)$

11. $(5x - 2y)(2x + 5y)$

12. $(-2a - 3b)(-5a + 2b)$

13. $(a - 2b)(b + 4x)$

14. $(4ab - c)(5ab + c)$

15. $(x + 2y)(-3x - y)$

16. $(7m - 2n)(3n - 5m)$

17. $(3x - 4y)(3x + 4y)$

18. $(1 + 2p)(3 - 4p)$

19. $(2a - 5b)(6b + 5a)$

20. $(7 - 5mp)^2$

21. $(5x + 3)(2x - 5)$

22. $(c - 2d)(3c + d)$

23. $(a + 3b)(-2a - b)$

24. $(5r + 2s)(3s + 2r)$

25. $(4a + 5c)^2$

26. $(e + f)(e - f)$

27. $(-c + 3d)(2c + 3d)$

28. $(6 + 5x)(-4x + 3)$

29. $(6x - 5y)(7x + y)$

30. $(3x - 7y)^2$

86

LESSON 33 – Multiplication of Polynomials

First, let's review multiplication of numbers.

$$
\begin{array}{r}
345 \\
\times\ 23 \\
\hline
1035 \\
690 \\
\hline
7935
\end{array}
\qquad\qquad
\begin{array}{r}
125 \\
\times\ 47 \\
\hline
875 \\
500 \\
\hline
5875
\end{array}
$$

There are two distinct methods of multiplication of polynomials: vertical or horizontal.

VERTICAL

Example 1: $(3x^2 + 4x - 5)(2x - 3)$

Solution:

Multiply by –3 first.

	$3x^2$	+	$4x$	–	5
			$2x$	–	3
	– $9x^2$	–	$12x$	+	15
$6x^3$ +	$8x^2$	–	$10x$		
$6x^3$ –	x^2	–	$22x$	+	15

Multiply by $2x$ first.

	$3x^2$	+	$4x$	–	3
			$2x$	–	3
$6x^3$ +	$8x^2$	–	$10x$		
	– $9x^2$	–	$12x$	+	15
$6x^3$ –	x^2	–	$22x$	+	15

Example 2: $(6a^2 - 3a + 2)(-a + 3)$

Solution:

	$6a^2$	–	$3a$	+	2
		–	a	+	3
	$18a^2$	–	$9a$	+	6
$-6a^3$ +	$3a^2$	–	$2a$		
$-6a^3$ +	$21a^2$	–	$11a$	+	6

HORIZONTAL

Example 3: $(3x^2 + 4x - 5)(2x - 3)$

Solution:

Rearrange: $(2x - 3)(3x^2 + 4x - 5)$
Use the Distributive Property: $2x(3x^2 + 4x - 5) = 6x^3 + 8x^2 - 10x$ **line 1**
Use the Distributive Property: $-3(3x^2 + 4x - 5) = -9x^2 - 12x + 15$ **line 2**

Add line 1 and line 2:

horizontal:

$(6x^3 + 8x^2 - 10x) + (-9x^2 - 12x + 15) =$
$6x^3 + (8x^2 - 9x^2) + (-10x - 12x) + 15 =$
$6x^3 + (-x^2) + (-22x) + 15$

$(3x^2 + 4x - 5)(2x - 3) = 6x^3 - x^2 - 22x + 15$

or, vertical: be sure to line up the variables correctly

$$
\begin{array}{rrrrr}
6x^3 & + & 8x^2 & - & 10x \\
 & - & 9x^2 & - & 12x & + & 15 \\
\hline
6x^3 & - & x^2 & - & 22x & + & 15
\end{array}
$$

Example 4: $(-a + 3)(6a^2 - 3a + 2)$

Solution:
Use Distributive Property: $(-a)(6a^2 - 3a + 2) = -6a^3 + 3a^2 - 2a$ **line 1**
Use Distributive Property: $3(6a^2 - 3a + 2) = 18a^2 - 9a + 6$ **line 2**

Add line 1 and line 2: $-6a^3 + (3a^2 + 18a^2) + (-2a - 9a) + 6$

$(-a + 3)(6a^2 - 3a + 2) = -6a^3 + 21a^2 - 11a + 6$

Multiply. Select a method and try it with these selected problems. Hint: Use FOIL on problems #1, #2, and #10.

1. $(3x - 4)(3x - 1)$ 2. $(4x + 9)(5x - 7)$

3. $(x + 2)(x^2 + x + 4)$ 4. $(x + 5)(3x^2 + x + 3)$

5. $(x + 2)(2x^2 + 2x + 5)$ 6. $(x + 4)(4x^2 - 3x + 6)$

7. $(x + 5)(x^2 + 2x + 5)$ 8. $(a + 4)(a^3 + a^2 - 7)$

9. $(b + 3)(b^2 - b - 8)$ 10. $(2c - 3)(8c + 5)$

LESSON 34 – Division of a Polynomial by a Monomial Divisor

To divide, break up the polynomial into separate terms, each term being divided by the divisor. Then divide each term. Check out these examples:

Example 1:

$$\frac{15x^3 + 12x^2 + 6x - 21}{3} =$$

$$\frac{15x^3}{3} + \frac{12x^2}{3} + \frac{6x}{3} - \frac{21}{3} =$$

$$5x^3 + 4x^2 + 2x - 7$$

Example 2:

$$\frac{16x^4 + 24x^3 - 8x^2 + 12x - 20}{4} =$$

$$\frac{16x^4}{4} + \frac{24x^3}{4} - \frac{8x^2}{4} + \frac{12x}{4} - \frac{20}{4} =$$

$$4x^4 + 6x^3 - 2x^2 + 3x - 5$$

Example 3:

$$\frac{16x^3 + 14x^2 - 8x - 6}{2} =$$

$$\frac{16x^3}{2} + \frac{14x^2}{2} - \frac{8x}{2} - \frac{6}{2} =$$

$$8x^3 + 7x^2 - 4x - 3$$

Example 4:

$$\frac{24x^3 + 17x^2 - 12x + 30}{3} =$$

$$\frac{24x^3}{3} + \frac{17x^2}{3} - \frac{12x}{3} + \frac{30}{3} =$$

$$8x^3 + \frac{17x^2}{3} - 4x + 10$$

Example 5:

$$\frac{4x^3 - 16x^2 + 11x - 8}{2x} =$$

$$\frac{4x^3}{2x} - \frac{16x^2}{2x} + \frac{11x}{2x} - \frac{8}{2x} =$$

$$2x^2 - 8x + \frac{11}{2} - \frac{4}{x}$$

> **NOTE**
>
> **If the divisor does not divide evenly, keep the fractional answer.**

Now try these for practice. Be sure to write fractions in reduced form.

1. $(16x^3 + 8x^2 - 4x - 24) \div 4$

2. $\dfrac{8x^4 + 6x^3 + 4x^2 + 2x + 1}{2}$

3. $(32x^3 - 24x^2 + 8x - 16) \div 8$

4. $\dfrac{4x^3 + 16x^2 - 9x + 12}{4}$

5. $(25x^3 + 35x^2 - 15x + 10) \div 5$

6. $\dfrac{15x^4 + 12x^3 - 18x^2 + 9x - 3}{3}$

7. $(28x^3 + 14x^2 + 35x) \div (7x)$

8. $(16a^3 - 12a^2 + 20a + 10) \div 4$

9. $\dfrac{40x^4 - 30x^3 + 16x^2 + 12a - 6}{8}$ be careful

10. $(6a^2b^2 - 18ab + 12) \div (3ab)$

11. $(18z^4 - 36z^3 + 24z^2 - 12z + 6) \div 6$

12. $(27x^5 + 36x^4 + 18x^3 - 9x^2 + 6x) \div (9x^2)$

13. $(10x^5 - 25x^4 + 15x^3) \div (-5x^3)$

14. $(24x^3y^2 - 36x^2yz + 48x) \div (12xy)$

15. $(16a^4 - 8a^3 - 24a^2) \div (-4a^2)$

16. $(5x^3 + 7x^2 - 2x) \div (-x)$

17. $(-9r^2s^3 + 27r^3s^2 + 48r^4s) \div (-3r^2s)$

18. $(-30x^5y^2z - 20x^2z^3 + 10y^3z^5) \div (5x^2yz^2)$

19. $(60r^2s - 24rs^2) \div 12rs$

20. $(28x^3y^2z - 21x^2y^3z^2 + 14xy^4z^3 - 7y^5z^4) \div (7xyz)$

21. $(36x^4 - 27x^3 + 81x^2 - 9x) \div (3x)$

22. $(16x^3 - 12x^2 - 24x + 40) \div 4$

23. $(12a^3b^3c^3 - 20a^2b^2c^2 + 8abc) \div (4abc)$

LESSON 35 – Division of a Polynomial by a Binomial Divisor

We will now use a binomial (2 terms) as our divisor. It sets up a little different. Remember those two-digit divisors in elementary school. I'll show you the similarities.

Elementary school example: $4296 \div 24$

Solution:

```
            1 7 9
   24 | 4 2 9 6
      − 2 4
        1 8 9
      − 1 6 8
          2 1 6
        − 2 1 6
              0
```

Check:

```
   ×   1 7 9
           2 4
         7 1 6
     3 5 8
     4 2 9 6
```

Example 1: $(3x^3 - x^2 + x - 1) \div (x + 1)$

Solution: $(3x^3 - x^2 + x - 1) \div (x + 1) = 3x^2 - 4x + 5$

Step 1: $\dfrac{3x^3}{x} = 3x^2$. Place $3x^2$ above the x^2 term.

Step 2: Multiply $3x^2$ times $(x + 1)$. $3x^2(x + 1) = 3x^2 + 3x^2$.

Step 3: Subtract. Remember when subtracting, the signs change.

Step 4: $\dfrac{-4x^2}{x} = -4x$. Place $-4x$ above the x term.

Step 5: Multiply: $-4x(x + 1) = -4x^2 - 4x$

Step 6: Subtract.

Step 7: $\dfrac{5x}{x} = 5$. Place above the constant (5).

Step 8: Multiply: $5(x + 1) = 5x + 5$

Step 9: Subtract.

Step 10: Check your answer.

```
                Step      Step      Step
                 1         4         7
                3x²   −   4x    +    5
  x + 1 |  3x³  −   x²   +    x   +   5
        −  3x³  −  3x²                         Step 3
             −  4x²   +   x
             +  4x²   +  4x                    Step 6
                    +  5x   +  5
                    −  5x   −  5               Step 9
                              0        remainder
```

```
Step 10:
              3x²   −   4x   +  5
  ×                           x  +  1
              3x²   −   4x   +  5
  3x³   −    4x²   +  5x
  3x³   −     x²   +   x   +  5

Checked!
```

Example 2: $(3x^3 - x^2 - 6x - 12) \div (x - 2)$

Solution: $(3x^3 - x^2 - 6x - 12) \div (x - 2) = 3x^2 + 5x + 4 + \dfrac{-4}{x-2}$

Step 1: $\dfrac{3x^3}{x} = 3x^2$. Place $3x^2$ above the x^2 term.

Step 2: Multiply: $3x^2(x - 2) = 3x^3 - 6x^2$.

Step 3: Subtract. Remember when subtracting, the signs change.

Step 4: $\dfrac{5x^2}{x} = 5x$. Place $5x$ above the x term.

Step 5: Multiply: $5x(x - 2) = 5x^2 - 10x$

Step 6: Subtract.

Step 7: $\dfrac{4x}{x} = 4$. Place above the constant (12).

Step 8: Multiply: $4(x - 2) = 4x - 8$

Step 9: Subtract. In this problem, there is a remainder.

Step 10: Check your answer.

$$
\begin{array}{r}
\ \ \ 3x^2\ +\ 5x\ +\ 4 \\
\hline
x-2\,\big|\ \ 3x^3\ -\ x^2\ -\ 6x\ -\ 12 \\
\underline{-\ 3x^3\ +\ 6x^2} \\
+\ 5x^2\ -\ 6x \\
\underline{-\ 5x^2\ +\ 10x} \\
+\ 4x\ -\ 12 \\
\underline{-\ 4x\ +\ 8} \\
-\ 4
\end{array}
$$

Step 1 (3x²), Step 4 (5x), Step 7 (4) labels above.
Step 3, Step 6, Step 9 (remainder) labels at right.

Step 10:

$$
\begin{array}{r}
3x^2\ +\ 5x\ +\ 4 \\
\times\ \ x\ -\ 2 \\
\hline
-\ 6x^2\ -\ 10x\ -\ 8 \\
3x^3\ +\ 5x^2\ +\ 4x \\
\hline
3x^3\ -\ x^2\ -\ 6x\ -\ 8 \\
-\ 4 \\
\hline
3x^3\ -\ x^2\ -\ 6x\ -\ 12
\end{array}
$$

Checked!

Make certain the remainder is a fraction using the divisor as the denominator.

Example 3: $(-x^3 - 4x - 9) \div (x - 2)$

Solution: $(-x^3 - 4x - 9) \div (x - 2) = -x^2 - 2x - 8 + \dfrac{-25}{x - 2}$

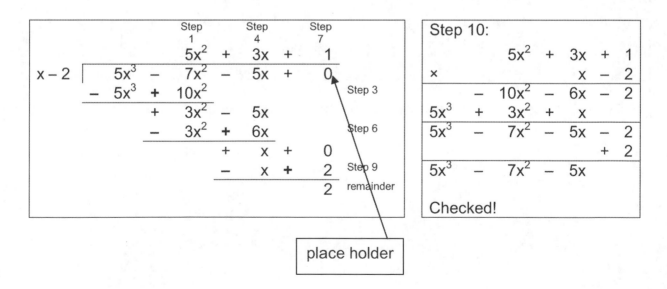

Step 1   Step 4   Step 7

$$
\begin{array}{r}
-\ x^2\ -\ 2x\ -\ 8 \\
x - 2\ \overline{)\ -\ x^3\ +\ 0\ -\ 4x\ -\ 9} \\
+\ x^3\ -\ 2x^2 \quad\quad\quad \text{Step 3}\\
\overline{\ -\ 2x^2\ -\ 4x} \\
+\ 2x^2\ -\ 4x \quad\quad \text{Step 6}\\
\overline{\ -\ 8x\ -\ 9} \\
+\ 8x\ -\ 16 \quad\quad \text{Step 9}\\
\overline{\ -\ 25} \quad \text{remainder}
\end{array}
$$

need place holder for x^2 position

Step 10:

$$
\begin{array}{r}
-x^2\ -\ 2x\ -\ 8 \\
\times \quad\quad x\ -\ 2 \\
\hline
2x^2\ +\ 4x\ +\ 16 \\
-x^3 \quad 2x^2\ -\ 8x \\
\hline
-x^3\ -\quad -\ 4x\ +\ 16 \\
-\ 25 \\
\hline
-x^3 \quad\quad -\ 4x\ -\ 9
\end{array}
$$

Checked!

Example 4: $(5x^3 - 7x^2 - 5x) \div (x - 2)$

Solution: $(5x^3 - 7x^2 - 5x) \div (x - 2) = 5x^2 + 3x + 1 + \dfrac{2}{x - 2}$

Step 1   Step 4   Step 7

$$
\begin{array}{r}
5x^2\ +\ 3x\ +\ 1 \\
x - 2\ \overline{)\ 5x^3\ -\ 7x^2\ -\ 5x\ +\ 0} \\
-\ 5x^3\ +\ 10x^2 \quad\quad\quad \text{Step 3}\\
\overline{\ +\ 3x^2\ -\ 5x} \\
-\ 3x^2\ +\ 6x \quad\quad \text{Step 6}\\
\overline{\ +\ x\ +\ 0} \\
-\ x\ +\ 2 \quad\quad \text{Step 9}\\
\overline{\ 2} \quad \text{remainder}
\end{array}
$$

place holder

Step 10:

$$
\begin{array}{r}
5x^2\ +\ 3x\ +\ 1 \\
\times \quad\quad x\ -\ 2 \\
\hline
-\ 10x^2\ -\ 6x\ -\ 2 \\
5x^3\ +\ 3x^2\ +\ x \\
\hline
5x^3\ -\ 7x^2\ -\ 5x\ -\ 2 \\
+\ 2 \\
\hline
5x^3\ -\ 7x^2\ -\ 5x
\end{array}
$$

Checked!

Example 5: $(x^3 + 125) \div (x + 5)$

Solution: $(x^3 + 125) \div (x + 5) = x^2 - 5x + 25$

Step 1 / Step 4 / Step 7

$$
\begin{array}{r}
x^2 - 5x + 25 \\
x + 5 \;\big)\; x^3 + 0 + 0 + 125 \\
- x^3 - 5x^2 \\
\hline
- 5x^2 + 0 \\
+ 5x^2 + 25x \\
\hline
+ 25x + 125 \\
- 25x - 125 \\
\hline
0
\end{array}
$$

Step 3

Step 6

Step 9

remainder

Step 10:

$$
\begin{array}{r}
x^2 - 5x + 25 \\
\times \qquad\quad x + 5 \\
\hline
5x^2 - 25x + 125 \\
x^3 - 5x^2 + 25x \\
\hline
x^3 \qquad\qquad + 125
\end{array}
$$

Checked!

place holders

Practice. Write any remainders as a fraction with the divisor as the denominator.

1. $(4x^3 - x^2 - 12x - 65) \div (x - 3)$

2. $(-x^3 + 4x + 6) \div (x + 1)$

3. $(z^3 - 729) \div (z - 9)$

4. $(-2x^3 - 3x^2 - 4) \div (x + 2)$

5. $(3x^3 + 7x^2 + 7x + 6) \div (x + 1)$

6. $(2x^3 + 5x^2 - 3x - 6) \div (x + 1)$

7. $(3x^3 - 19x^2 + 27x + 4) \div (x - 4)$

8. $(2y^2 - 11y + 12y^3 + 3) \div (y - 3)$

9. $(a^3 - 64) \div (a - 4)$

10. $(2x^2 - 7x - 8) \div (2x - 1)$

11. $(8x^3 - 12x^2 + 6x - 1) \div (2x - 1)$

12. $(x^3 - x^2 - 4x + 4) \div (x + 2)$

LESSON 36 – A Review of Lessons 30-35

Perform the indicated operation(s). Write answers in descending order of the variable and the remainders as fractions.

1. $(-7x^2 - 5x^3 + 3) - (3x^3 + 7 + 5x^2)$

2. $(-5x^3 - 5x^2 - 7x - 9) + (-3x^3 + 5x^2 - 3)$

3. $(-5x^4 + 5x^3 - 9) + 3(9x^4 + 7x^3 + x + 5)$

4. $2(-3x^5 - x^4 - 2x^3 - 6x^2 - 5x + 4) + 3(-3x^4 - x^3 - 2x^2 - 6x - 5)$

5. $(4x^2 - 4x + 2) + (-3x^2 + 5x - 5) =$
    a. $x^2 + x - 3$ c. $x^2 - 9x + 7$
    b. $7x^2 + x + 7$ d. $7x^2 - 9x - 3$

6. $(5x^2 - 5)^2$

7. $(2x + 1)(3x + 7)$

8. $(x + 4)(x^2 - 4x - 1)$

9. $(x - 2)(3x^2 + x - 6) =$
    a. $3x^3 - 5x^2 - 8x + 12$ c. $3x^2 + 12$
    b. $3x^3 + x^2 + 12$ d. $3x^3 - 5x^2 - 2x + 12$

10. $(3x^3 - x^2 - 2x - 20) \div (x - 2)$

11. $(x^3 - 5x + 1) \div (x + 2)$

12. $(n^3 + 27) \div (n + 3)$

13. $(4x^3 - 6x^2 + 2) \div (x - 1)$

14. $(6x^3 - 2x^2 - 2x - 1) \div (x - 1)$

15. $(2x^3 - 2x + 1) \div (x + 2) =$
    a. $2x^2 - 6x - 12 + \dfrac{25}{x + 2}$

    b. $2x^2 - 6x + 13 - \dfrac{26}{x + 2}$

    c. $2x^2 - 4x - 10 + \dfrac{19}{x + 2}$

    d. $2x^2 - 4x + 6 - \dfrac{11}{x + 2}$

16. $3(-2x^5 + 5x^4 - 3x^3 - x^2 - 4x + 6) - 4(2x^4 - 5x^3 + 3x^2 + x - 4)$

17. $2(6x^5 - 2x^4 - 5x^3 + 3x^2 + x - 4) + 3(6x^4 + 2x^3 - 5x^2 - 3x - 1)$

18. $4(-x^5 - 4x^4 + 6x^3 - 2x^2 - 5x - 3) - 2(x^4 + 4x^3 + 6x^2 - 2x + 5)$

19. $3(-3x^5 + x^4 + 4x^3 + 6x^2 + 2x + 5) + 4(-3x^4 - x^3 + 4x^2 + 6x + 2)$

20. $(5x^2 - 3x + 4) + (-3x^2 + 3x + 3) =$
    a. $8x^2 - 6x + 7$
    b. $8x^2 + 1$

    c. $2x^2 - 6x + 1$
    d. $2x^2 + 7$

21. $(x + 4)(x^2 - 2x + 3)$

22. $(x + 3)(x^2 + 3x + 1)$

23. $(x + 3)(2x^2 - 3x - 1) =$
    a. $2x^3 + 3x^2 - 10x - 3$
    b. $2x^2 - 4x - 3$

    c. $2x^3 + 3x^2 - 9x - 3$
    d. $2x^3 - 3x^2 - 3$

24. $6(x - 2)(x + 4)$

25. $\dfrac{16y^3 - 8y^2 + 6y - 10}{2y}$

96

LESSON 37 – Common Monomial Factor (CMF)

Factoring, in algebra, requires an ability to recognize special situations along with knowing prime numbers. Prime numbers, by definition, have itself and one (1) as its only two factors. Single digit prime numbers are 2, 3, 5, and 7. With these concepts, let us begin to understand factoring.

We will begin with the common monomial factor (CMF). Look for this first and foremost when factoring.

Let us examine the five pairs of factors for 48. They are: (1)(48), (2)(24), (3)(16), (4)(12), and (6)(8). How can factoring fundamentals help in math? It helps when trying to "factor" a binomial (2 terms), a trinomial (3 terms), and a polynomial (more than 3 terms).

Let us begin by factoring a binomial. That is two terms. Examples of a term: x, 3x, $4x^2$.

Example: 2x + 6b

Solution:

2 is a factor of itself and also 6. $\dfrac{2x}{2} + \dfrac{6b}{2} = x + 3b$

When we take a 2 out of 2x + 6b, we end up with 2(x + 3b).

This mathematics process of factoring is the removal of a common monomial (one term) factor. From the above example, I essentially took out a 2 from 2x, leaving an x, and took out a 2 from 6b, leaving 3b. The final result: 2x + 6b = 2(x + 3b).

More binomial factoring problems to practice:

a. $16x^2 + 64y^2$
b. $4a^3b^2c^4 - 18a^2b^3c^3$
c. $9x^2 - 25y^2$
d. $8 + 24a^2$
e. $12x^4y^2 - 32a^2x^4$

Several trinomials with common monomial factor:

f. $2x^2 + 6x + 10$
g. $3a^2x + 6ax - 12x$

Let's examine each situation.

a. What common factor is present in $16x^2 + 64y^2$? Certainly not the x^2 nor y^2. Okay, let's look at 16 and 64. This is where factoring knowledge comes in handy.

Factors of 16 are: 1 and **16**, 2 and 8, 4 and 4
Factors of 64 are: 1 and 64, 2 and 32, 4 and **16**, 8 and 8

The largest common factor of both 16 and 64 is 16. Therefore, 16 is the common monomial factor (CMF). Removing the 16 from both terms, ***and*** placing it outside the parenthesis, our answer is $16(x^2 + 4y^2)$.

b. What common factor is present in $4a^3b^2c^4 - 18a^2b^3c^3$? At least a 2, a, b, and c. How many a's, b's, and c's? Pick a minimum of each term. For a, that would be a^2; for b, it would be b^2; and for c, it's c^3. Therefore, the CMF would be $2a^2b^2c^3$.

Removing the CMF from the original problem, what remains? In the first term it is 2ac and in the second terms, 9b.

The final answer is $2a^2b^2c^3(2ac - 9b)$.

c. There is no CMF in $9x^2 - 25y^2$. However, this binomial is an example of a perfect square term minus another perfect square term. One needs to take the square root of each term, 3x and 5y, and have opposite signs for each factor. The final answer is $(3x + 5y)(3x - 5y)$. More about square minus square, cube plus cube, and cube minus cube in Lesson 38.

d. $8 + 24a^2$ has a CMF in each term. The magic CMF is 8. When 8 is removed from both terms, we end up with 1 and 3a2. Yes, we need a 1 to replace the 8. So the correct answer is 8(1 + 3a2).

e. $12x^4y^2 - 32a^2x^4$ has a CMF of $4x^4$. Removing the $4x^4$ from both terms leaves $3y^2 - 8a^2$. The final answer is $4x^4(3y^2 - 8a^2)$.

f. In $2x^2 + 6x + 10$, 2 is the CMF in all three terms. Completely factored, $2(x^2 + 3x + 5)$ is the final answer.

g. Factoring a 3x from the original problem, $3a^2x + 6ax - 12x$, we get $3x(a^2 + 2a - 6)$ which is fully factored.

Some practice problems to try. Factor by removing the CMF

1. $2x^2 + 4x^3$

2. $16a^2 + 32b^2$

3. $4x^2 + 6xy + 8y^2$

4. $3x^4 + 12x^3 + 18x^2$

5. $4x^2 + 16x^3 + 32x$

6. $5x^2 + 15x^3 + 25x^4$

7. $81y^3 - 36xy^2$

8. $9a^3b^3 + 3a^2b^2 + 6a^4b^4$

9. $8y^3z^3 + 56y^4z^3$

10. $60a^2b^3c^4 - 24a^3b^2c^3$

11. $18x^2y^2z^2 + 20xyz$

12. $16y^2 + 32y + 16$

13. $12x^5y^3 - 16x^2y^2$

14. $28y^2 - 20y + 16y^3$

15. $48x^4y^3z^2 - 6x^2y^3z^4$

16. $49x^2 - 63y^3$

17. $7x^2y^3 + 21x^3y^4 + 28x^4y^5$

18. $54m^2np^3 + 27mn^2p^2 - 18m^3p^2$

19. $9a^3b^2c^2 - 18abc^2 + 12a^2b - 30ab^2c^2$

20. $16ef^2g^3 - 8e^2f^2g^2 + 4e^2fg^2 - 20e^3f^2g^3$

21. $14x^3y^2z^4 - 28x^2y^3z^2 + 35x^5y^2z^3 - 49x^2y^4z^5$

22. $15a^3bc^2 + 25a^2b^2c^2 - 30a^4b^3c^4 + 60a^3b^2c^4$

23. $18L^2m^3n^4 - 36L^3m^3n^3 + 24L^4m^4n^5 + 48L^3m^2n^3$

24. $21r^2s^3t^4 + 30r^3s^2t^2 + 6r^2s^4t^5 + 18r^4s^2t^4$

LESSON 38 – Factoring a Binomial
It could be either a square minus a square,
a cube plus a cube, or a cube minus a cube.

First, one needs to know perfect square numerals. Here's some help. 1, 4, 9, 16, 25, 36, 49, 64, 81, and 100 are the first ten numbers squared. These squares are numbers to know. Then the cubes. 1, 8, 27, 64, 125, 216, 343, 512, 729, and 1000 are the first ten cubes. Wow! When you have memorized the above facts, we can proceed. Get that calculator working.

Here are some practice problems with answers:

Example 1: $27x^3 + 64y^6$ cube plus cube

> Solution:
>
> There is a formula for cubes. Here's the scoop:
>
> $$a^3 + b^3 = (a + b)(a^2 - ab + b^2)$$
> $$a^3 - b^3 = (a - b)(a^2 + ab + b^2)$$
>
> So let's try to factor $27x^3 + 64y^6$. The cube root of $27x^3$ is $3x$ and the cube root of $64y^6$ is $4y^2$. So the first factor of the answer is $(3x + 4y^2)$. Now let's figure out the second factor. Using the factor $(3x + 4y^2)$, square the first term $(3x)$, which is $9x^2$. Then, multiply the two terms together, $(3x)(4y^2)$, which is $12xy^2$. Finally, square the second term $(4y^2)$ for $16y^4$. So, putting this altogether, the final answer is $(3x + 4y^2)(9x^2 - 12xy^2 + 16y^4)$.

Example 2: $8a^3 - 125b^3$ cube minus cube

> Solution:
>
> Take the cube root of $8a^3$ and $125b^3$, getting $2a$ and $5b$, respectively. The first factor is $(2a - 5b)$. Use this factor to arrive at the second factor: square the first term, multiply the two terms together, and square the second term: $(2a)^2 + (2a)(5b) + (5b)^2 = 4a^2 + 10ab + 25b^2$. Putting the two factors together to get $(2a - 5b)(4a^2 + 10ab + 5b^2)$, which is the correct answer.

Example 3: $100x^4 - 25y^2$ square minus square $\boxed{} - \boxed{}$

Solution:

In Lesson 37, practice problem c, it was a square minus square factor problem. Again we have a square minus square problem. We must first check for a CMF and, low and behold, there is one. The CMF is 25. Removing the CMF, we arrive at $25(4x^4 - y^2)$. The second factor $(4x^4 - y^2)$ is a square minus square and needs to be factored further. We need to take the square root of each term, resulting in $(2x^2 + y)(2x^2 - y)$. The final answer is $25(2x^2 + y)(2x^2 - y)$.

Remember the formulas!

$$a^2 - b^2 = (a + b)(a - b)$$
$$a^3 + b^3 = (a + b)(a^2 - ab + b^2)$$
$$a^3 - b^3 = (a - b)(a^2 + ab + b^2)$$
$\left.\right\}$ Remember these!

Also, L⊙⊙K for a common monomial factor every time you factor.

So far for factoring:

1.   Common monomial factor (CMF)
2.   If a binomial (2 terms):
         square minus square
         cube plus cube
         cube minus square

Now try these problems. Factor the following.

1. $a^2 - 36$ 2. $4x^2 - 36$

3. $a^2b^2 + a^3b^3$ 4. $a^3 - 64$

5. $9x^2 - 81y^2$ 6. $8b^3 + 27$

7. $x^2y^2 - a^2$ 8. $64x^3 - 125$

9. $16e^2 - 49f^2$

10. $16e^2 - 36f^2$

11. $c^3d^3 + 1$

12. $1 - 4a^2$

13. $100x^2 - 25$

14. $1 + 8z^3$

15. $81x^4 - 9a^2x^2$

16. $16a^2 - 64b^2$

17. $y^3 - 216x^3$

18. $125x^3 + 64y^6$

19. $b^2 - 9a^2$

20. $343m^6 - 8n^3$

21. $1 + 64x^3$

22. $64y^6 - 512z^3$

23. $49y^2 - 4z^2$

24. $8a^3 + 125$

25. $1000 - 125p^6$

26. $144a^2 - 64b^2$

27. $125c^3 + 64d^3$

28. $216x^6 - 343y^9$

29. $64x^6 - 1$

30. $1 - 169x^4$

LESSON 39 – Factoring Four Terms by Grouping

In the previous two lessons, we looked for a common monomial factor (CMF) in a group of terms and we examined binomials (two terms. Binomials are basically (i) square minus square, (ii) cube plus cube, and (iii) cube minus cube.

Now we will look at polynomials (many terms). The first group of polynomials is four terms. When we see four terms, we should think "grouping".

So to summarize to this point, we follow these guidelines:

1. CMF – common monomial factor
2. 2 terms – binomials (square minus square, cube plus cube, cube minus cube)
3. 4 terms – grouping

Here is an example for factoring by grouping and how it is factored.

Example 1: $8a^2 - 10ab + 12ab - 15b^2$

Solution:

Step 1: Group the first two terms and the last two terms.

$$(8a^2 - 10ab) + (12ab - 15b^2)$$

Step 2: Find a CMF in $8a^2 - 10ab$. The CMF is 2a. Now remove, by dividing, the 2a from both terms, which leaves: $2a(4a - 5b)$. With this idea in mind, find a CMF in $12ab - 15b^2$. It is 3b. When dividing 3b into $12ab - 15b^2$, the answer is $4a - 5b$. So factored, it is $3b(4a - 5b)$.

Let's review what we have so far:

$$8a^2 - 10ab + 12ab - 15b^2 =$$
$$(8a^2 - 10ab) + (12ab - 15b^2) =$$
$$2a(4a - 5b) + 3b(4a - 5b).$$

Now there is a CMF in both factored terms. It is $4a - 5b$. Extract $4a - 5b$ from both terms. What remains is $2a + 3b$.

Completely factored, $8a^2 - 10ab + 12ab - 15b^2 = (4a - 5b)(2a + 3b)$.

Using the FOIL method, let's check our work.

First   Outer   Inner   Last

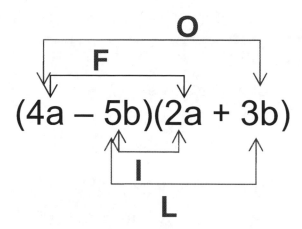

First terms: $4a \cdot 2a = 8a^2$
Outer terms: $4a \cdot 3b = 12ab$
Inner terms: $-5b \cdot 2a = -10ab$
Last terms: $-5b \cdot 3b = -15b^2$

$$8a^2 + 12ab - 10ab - 15b^2$$

Example 2: $2x^2 - 10x + 3xy - 15y$

Solution:

Step 1:   Group first two terms and last two terms.

$$(2x^2 - 10x) + (3xy - 15y)$$

Step 2:   Find a CMF in both groups. $2x^2 - 10x$ have $2x$ in common and $3xy - 15y$ have $3y$ in common.

Step 3: $\dfrac{2x^2 - 10x}{2x} = x - 5$ $(2x^2 - 10x)$ factored is $2x(x - 5)$

$\dfrac{3xy - 15y}{3y} = x - 5$ $(3xy - 15y)$ factored is $3y(x - 5)$

So far, we have

$$2x^2 - 10x + 3xy - 15y = 2x(x - 5) + 3y(x - 5)$$

Step 4: Looking at $2x(x-5) + 3y(x-5)$, we have another CMF $(x-5)$. Removing it, we arrive at the two factors of $2x^2 - 10x + 3xy - 15y$.

$$2x(x-5) + 3y(x-5) = (x-5)(2x+3y)$$

Therefore, $2x^2 - 10x + 3xy - 15y = (x-5)(2x+3y)$

Step 5: Check the answer by FOIL.

$$
\begin{array}{ll}
\text{F:} & x \cdot 2x = 2x^2 \\
\text{O:} & x \cdot 3y = 3xy \\
\text{I:} & -5 \cdot 2x = -10x \\
\text{L:} & -5 \cdot 3y = -15y
\end{array}
$$

$$2x^2 + 3xy - 10x - 15y$$
$$\text{or}$$
$$2x^2 - 10x + 3xy - 15y$$

Example 3: $3x^2y - 12x^2 + 9xy - 36x$

Solution:

Step 1: Remove the CMF: $3x$

$$3x(xy - 4x + 3y - 12)$$

Step 2: Group the first two terms and the last two terms.

$$3x[(xy - 4x) + (3y - 12)]$$

Step 3: Remove "x" from the first two terms and "3" from the last two terms.

$$3x[x(y-4) + 3(y-4)]$$

Step 4: Remove the CMF "y – 4".

$$3x[(y-4)(x+3)]$$
$$3x(y-4)(x+3)$$

Step 5: Check by FOIL.

Solution: $3x(y-4)(x+3)$

Example 4: $x^2 - 21y - 7x + 3xy$

        Solution:

                Step 1:   Rearrange the order of the terms: $x^2 - 7x + 3xy - 21y$

                Step 2:   Group: $(x^2 - 7x) + (3xy - 21y)$

                Step 3:   Remove CMF's: $x(x - 7) + 3y(x - 7)$

                Step 4:   Remove CMF: $(x - 7)(x + 3y)$

                Step 5:   Check by FOIL.

        Solution: $(x - 7)(x + 3y)$

Example 5: $-b - a + 1 + ab$

        Solution:

                Step 1:   Rearrange order: $ab - b - a + 1$

                Step 2:   Insert parenthesis: $ab - b - (a - 1)$ <u>it's ok</u>

                Step 3:   Group: $(ab - b) - (a - 1)$

                Step 4:   Remove CMF's: $b(a - 1) - 1(a - 1)$

                Step 5:   Remove CMF: $(a - 1)(b - 1)$

                Step 6:   Check by FOIL.

        Solution: $(a - 1)(b - 1)$

Let's review several problems about grouping.

1. $(a + b)(2 + c) + x(2 + c)$ 2. $x^2(y + 2) - 5(y + 2)$

3. $x(x - 2) - 6(x - 2)$ 4. $pq + 5q + 2p + 10$

5. $2xy + 3y + 2x + 3$ 6. $2a^2 - 4a + 3ab - 6b$

7. $x^3 + 3x^2 - 5x - 15$

8. $6ax + 24x + a + 4$

9. $ax^2 - 5y^2 + ay^2 - 5x^2$
   hint: rearrange the order

10. $15 - 5y - 3x + xy$

11. $xy + 3y + 2x + 6$

12. $a^2b + ab^2c + ca + bc^2$

13. $2ab - 7a + a^2 - 14b$

14. $x^2 - 2y + 2x - xy$

15. $ap + bp - ax - bx$

16. $3x^2y - 12x^2 - 12x^3 + 3xy$

17. $10xy^2 - 5y^2 + 30xy - 15y$

18. $ab^3 + b^4 + 6ab^2 + 6b^3$

19. $8 - 4y^2 + 8x - 4xy^2$

20. $x^2y - rs - x^2s + ry$

21. $y^3 + 2y^2 - 3y - 6$

22. $9ab - 4 + 12a - 3b$

23. $x^3 + x^2 + 9x + 9$

24. $2c^2 + 20d - 8c - 5cd$

25. $2xy + 12 - 3y - 8x$

26. $12ab + 18a^2 - 3ac - 2bc$

27. $5x^2 + 15xy - 2xz - 6yz$

28. $xy^2 - 2y^2 - x + 2$

LESSON 40 – Review of Factoring
CMF, Binominals, and Grouping

Review factoring with these concepts in mind:

#1:   Look for a CMF.
#2:   If it's a binomial, check for square minus square, cube plus cube, or cube minus cube.
#3:   Grouping with a four-term polynomial.

Now we'll practice to see if we can move forward.

Example 1: $x^2(y + 2) - 9(y + 2)$

    Solution:

        Find a CMF if present. The CMF is $(y + 2)$. Remove the CMF from both terms, arriving at $(y + 2)(x^2 - 9)$. ARE WE DONE?

        NO! Why not? Correct, you can STILL factor the $(y^2 - 9)$. It is a square minus a square. $(x^2 - 9)$ factors into $(x + 3)$ and $(x - 3)$.

        Result: $(y + 2)(x + 3)(x - 3)$

Example 2: $x^4 + x^3 + 8a^3x + 8a^3$

    Solution:

        Group the first two terms and the last two terms:

$$(x^4 + x^3) + (8a^3x + 8a^3)$$

        L⊙⊙K for a CMF in both groups and remove it.

        $(x^4 + x^3) = x^3(x + 1)$
        $(8a^3x + 8a^3) = 8a^3(x + 1)$

        We now have: $x^3(x + 1) + 8a^3(x + 1)$.

        Again, look for a CMF in both groups and remove it, $(x + 1)$ is common to both. Now we have: $(x + 1)(x^3 + 8a^3)$. But $(x^3 + 8a^3)$ is a cube plus cube. It factors into $(x + 2a)(x^2 - 2ax + 4a^2)$.

        Final answer: $(x + 1)(x + 2a)(x^2 - 2ax + 4a^2)$.

Now we are handling the factoring dilemmas.

Example 3: $x^4 - 16x^2$

> Solution:
>> We have a CMF: x^2.
>>
>> Removing x^2 from both terms we get: $x^2(x^2 - 16)$. Is this fully-factored? No. $(x^2 - 16)$ is a square minus a square. It factors into $(x + 4)$ and $(x - 4)$.
>>
>> Final answer: $x^2(x + 4)(x - 4)$

Example 4: $x^3(x^2 - 9) - y^3(x^2 - 9)$

> Solution:
>> Double trouble!!
>>
>> First factor out the CMF of $(x^2 - 9)$. What remains is $(x^3 - y^3)$. Giving us $(x^2 - 9)(x^3 - y^3)$. Both of which need factoring again.
>>
>> $(x^2 - 9)$ is a square minus square. $(x^3 - y^3)$ is a cube minus cube.
>>
>>> $(x^2 - 9) = (x + 3)(x - 3)$
>>> $(x^3 - y^3) = (x - y)(x^2 + xy + y^2)$
>>
>> Putting it altogether, we have: $(x + 3)(x - 3)(x - y)(x^2 + xy + y^2)$

Example 5: $16x^4 - 256y^4$

> Solution:
>> Remove the CMF: $16(x^4 - 16y^4)$.
>>
>> Factor $(x^4 - 16y^4)$, a square minus square: $(x^2 + 4y^2)(x^2 - 4y^2)$.
>>
>> Factor $(x^2 - 4y^2)$, another square minus square: $(x + 2y)(x - 2y)$.
>>
>> Finally we have our answer: $16(x^2 + 4y^2)(x + 2y)(x - 2y)$.
>>
>> That was lots of work!

One more for good measure!

Example 6: $a^4x^4 - 81b^4y^4$

Solution:

There is no CMF. But it is a square minus square.

$$a^4x^4 - 81b^4y^4 = (a^2x^2 + 9b^2y^2)(a^2x^2 - 9b^2y^2)$$

However $(a^2x^2 - 9b^2y^2)$ is another square minus square.

$$(a^2x^2 - 9b^2y^2) = (ax + 3by)(ax - 3by)$$

Fully factored we have: $(a^2x^2 + 9b^2y^2)(ax + 3by)(ax - 3by)$

Try these few problems. Watch out for any "double trouble"!!

1. $20x^2 - 8xy + 15xy - 6y^2$ 2. $8x^2 - 20x + 10x - 25$

3. $s^2 + 4s + 7s + 28$ 4. $8x^2 - 20xy - 6xy + 15y^2$

5. $8x^3 + 27y^6$ 6. $16x^2 - 25y^2$

7. $x^2y^3 + 8x^2z^3$ 8. $64 - x^6$

9. $(x^2 - y^2) - z^2(x^2 - y^2)$ 10. $16x - x^3$

11. $25x^4 - 225y^6$ 12. $16a^3 - 4a^2b^2 - 4ab + b^3$

LESSON 41 – Factoring a Trinomial When the Coefficient of x^2 is 1
(Trial and Error)

Factoring a trinomial whose coefficient of the squared variable is +1, we can use the trial and error theory.

Example 1: $x^2 + 4x + 3$

Solution:

Step 1:   (x +   )(x +   )   set up this first
Step 2:   (x + 3)(x + 1)   only factors of 3 are (1)(3)
Step 3:   Check it using FOIL!

$$F \rightarrow x^2$$
$$\boxed{\begin{array}{l} O \rightarrow x \\ I \rightarrow 3x \end{array}} = 4x$$
$$L \rightarrow 3$$

$$x^2 + 4x + 3 \quad \text{GOOD!}$$

Example 2: $x^2 - 4x + 3$

Solution: Notice that this is the same as example 1 except for the signs.

Step 1:   (x −   )(x −   )
Step 2:   (x − 3)(x − 1)
Step 3:   Remember to check by using FOIL!

Now we have two locks!

#1: $x^2 + ax + b$, where both signs are positive, must always be (x +   )(x +   ).

#2: $x^2 - ax + b$, where first sign is negative and second sign is positive, must always be (x −   )(x −   ).

However, if we have a two negative-sign trinomial or a positive-negative-sign trinomial, the signs of the factors will be one positive sign and one negative sign. The sign of the middle term will determine the largest factors' sign.

Check out the following examples.

Example 3: $x^2 - 5x - 24$

Solution:
This will factor into $(x - 8)(x + 3)$. Let me explain. The last term is 24 which has several factors: (24)(1), (12)(2), (8)(3), and (6)(4).

(24)(1):   gives 25 when added   gives 23 when subtracted
(12)(2):   gives 14 when added   gives 10 when subtracted
(8)(3):   gives 11 when added   gives 5 when subtracted
(6)(4):   gives 10 when added   gives 2 when subtracted

In the original trinomial, the middle term is –5x. Therefore we need the factors 8 and 3, where 8 gets the negative sign and 3 gets the positive sign. Therefore, $x^2 - 5x - 24$ factored is $(x - 8)(x +3)$. Check using FOIL.

Example 4: $x^2 + 5x - 24$

Same as above example but the signs are reversed. Since we now have a +5x as the middle term the factors are $(x + 8)(x - 3)$. Use FOIL to check!

Now try these practice problems. Mark this page so you can refer to it easily.

1. $a^2 - 2a - 15$ 2. $x^2 + 5x - 36$

3. $b^2 + 4b - 21$ 4. $x^2 - 2x - 35$

5. $x^2 + 11x - 26$ 6. $c^2 + 3c - 40$

7. $r^2 + 4r - 96$ 8. $m^2 + 3m - 54$

9. $y^2 + 3y - 28$ 10. $e^2 - 2e - 99$

11. $f^2 + 11f - 80$ 12. $a^2 - 9a - 36$

13. $x^2 - 2x - 48$ 14. $x^2 - 5x - 50$

15. $x^2 + 7x - 60$ 16. $e^2 - 17e + 72$

LESSON 42 – Factoring a Trinomial When the Coefficient of $x^2 > 1$

This is the final installment on understanding how to factor. It's factoring a trinomial whose x^2 coefficient is greater than 1. It involves grouping, setting up four terms to factor. In general, it looks like this: $Ax^2 + Bx + C$.

Let me explain a few things before your try. Remember, we are looking for two factors, when multiplied, to give us the required trinomial. So here it goes!

Example 1: $2x^2 - 7x - 15$

    Solution:

        The key here is to multiply the x^2 coefficient and the constant. In the above trinomial, that would be the 2 and –15. The result is –30.

        Next we examine what two factors of –30 has a sum or difference of –7. (That's the coefficient of the middle term.) Our choices are: (1)(–30), (2)(–15), (3)(–10), (5)(–6), (6)(–5), (10)(–3), (15)(–2), or (30)(–1). From these eight choices, there is one obvious answer: (3)(–10). Having selected these two factors, we rewrite the trinomial and make it a four-term polynomial: (a) $2x^2 + 3x - 10x - 15$ or (b) $2x^2 - 10x + 3x - 15$. Either way, our answer will be the same.

        Let us try (b) first. Group the first two terms and the last two terms:

            $2x^2 - 10x + 3x - 15$
            $(2x^2 - 10x) + (3x - 15)$.

        Next, take a common factor out of both groups: 2x in the first group and 3 in the second group.

        Removing the 2x from the two terms in the first group we have: $2x(x - 5)$. Likewise, the 3 is removed from the two terms in the second group, resulting in $3(x - 5)$.

        Combining everything we have done so far, we have:

            $2x^2 - 10x + 3x - 15$
            $(2x^2 - 10x) + (3x - 15)$
            $2x(x - 5) + 3(x - 5)$

        From the last step, we show a CMF (common monomial factor) in both groups. It is $(x - 5)$. Remove the $(x - 5)$ from both groups. What remains is $(2x + 3)$. The final answer is $(x - 5)(2x + 3)$.

Check your work by multiplying $(x - 5)$ times $(2x + 3)$.

F $\quad$ $(x)(2x) = 2x^2$
O $\quad$ $(x)(+3) = +3x$
I $\quad$ $(-5)(2x) = -10x$
L $\quad$ $(-5)(+3) = -15$

Combine O $(+3x)$ and I $(-10x)$ to get $-7x$.

Final check: $2x^2 - 7x - 15$.

When we try (a) $2x^2 + 3x - 10x - 15$, we should get the same answer.

Separate the $(2x^2 + 3x)$ from the $(-10x - 15)$.

$(2x^2 + 3x) - (10x - 15)$
$(2x^2 + 3x) - (10x + 15)$

> This negative sign changes the -15 to $+15$.

Find a CMF in both groups and remove it. In $(2x^2 + 3x)$, it is "x". In $(10x + 15)$, it is "5". This results in:

$x(2x + 3) - 5(2x + 3)$

Remove the CMF from both groups, which is $(2x + 3)$.

$(2x + 3)(x - 5)$

NOTE: your can reverse the factors also: $(x - 5)(2x + 3)$. Therefore, the answer can be $(x - 5)(2x + 3)$ or $(2x + 3)(x - 5)$.

Example 2: $\quad 6x^2 + x - 35$.

Solution:
Product $(6 \cdot -35)$ is -210. Difference is 1.

Factors of -210 are: $(1)(-210)$, $(-1)(210)$, $(2)(-105)$, $(-2)(105)$, $(3)(-70)$, $(-3)(70)$, $(5)(-42)$, $(-5)(42)$, $(6)(-35)$, $(-6)(35)$, $(7)(-30)$, $(-7)(30)$, $(10)(-21)$, $(-10)(21)$, $(14)(-15)$, $(-14)(15)$. Our last try gets us a difference between factors of $+1$.

Rewrite the trinomial as a four-term polynomial.

$6x^2 - 14x + 15x - 35$

Group. (Note, put the negative as the second term and the positive as the third terms for grouping the four-term polynomial.)

$$(6x^2 - 14x) + (15x - 35)$$

Find the CMF for each and remove it.

$$2x(3x - 7) + 5(3x - 7)$$

Again, find the CMF and remove it.

$$(3x - 7)(2x + 5)$$

Check the final answer.

F: $(3x)(2x) = 6x^2$
O: $(3x)(5) = 15x$
I: $(-7)(2x) = -14x$
L: $(-7)(5) = -35$

Combining O (15x) and I (–14x) for the middle term of +x.

$$6x^2 + x - 35 \text{ voila!}$$

Example 3: $5x^2 - 31x - 28$.

Solution:
We know for sure that the factors need to look something like this (5x   )(x   ) because the only factors of 5 are itself and one.

Now examine –28. There are several factors besides itself and one. They are (14)(2) and (7)(4).

Taking the factors (14)(2), we multiply 14 by 5 gives 70, that will be 70 ± 2 or 72 and 68. Whereas the factors (7)(4) would be (7 times 5) ± 4, or 39 and 31. There is our 31.

Now substitute in 7 and 4 with their signs so (5x + 7)(x – 4) gives –13x for the middle term. Oops. Let's reverse the order: (5x + 4)(x – 7) gives –31x for the middle term. Therefore, (5x + 4)(x – 7) is what we want. Check it using FOIL.

Try the above method (called T & E) for these:

1. $7x^2 + 23x + 6$ 2. $2x^2 - 11x + 14$

3. $4x^2 + 17x - 15$ 4. $5x^2 + 13x - 6$

5. $3x^2 + 13x + 4$ 6. $5x^2 + 6x - 8$

7. $4y^2 - 11y + 6$ 8. $9x^2 - 21x + 10$

9. $6x^2 + 5x - 4$ 10. $6n^2 - 11n - 10$

11. $4x^2 - 3x - 7$ 12. $9a^2 - 18a + 8$

13. $10x^2 - 23x + 12$ 14. $6x^2 + x - 1$

15. $2x^2 + 7x + 6$ 16. $12x^2 + 11x - 5$

17. $20x^2 + 11x - 3$ 18. $15x^2 + x - 2$

19. $6x^2 - 17x + 12$ 20. $2b^2 + b - 28$

LESSON 43 – More Problems on Factoring Trinomials

1. Factor completely.
   a. $6x^2 - 6x - 36$

   b. $4x^5 + 20x^4 - 24x^3$

2. Find the pair of numbers whose product is 11 and whose sum is 12.

3. Use the FOIL method to show that $(9x + 18)(x - 10)$ is $9x^2 - 72x - 180$. If you where asked to completely factor $9x^2 - 72x - 180$, why would it be incorrect to give $(9x + 18)(x - 10)$ as your answer?

4. What steps would you take to factor $x^2 - 9x + 20$?

5. Factor completely: $x^2 - x - 56$

6. Find the pair of numbers whose product is 24 and whose sum is 11.

7. Factor completely.
   a. $x^2 + 2x - 120$ b. $3x^3 + 9x^2y - 30xy^2$ c. $2x^2 - 10x + 12$

8. In factoring a trinomial in "z" as $(z + a)(z + b)$, what must be true of "a" and "b", if the coefficient of the last term of the trinomial is positive?

9. Factor completely.
   a. $x^2 + 2xy - 99y^2$ b. $5x^3 + 10x^2 - 75x$ c. $x^2 - x - 45$

10. Explain the error in the following: $x^2 + 2x - 15 = (x - 5)(x + 3)$

11. Factor completely.
    a. $x^2 - 8x - 48$ b. $x^2 + 14x - 15$

12. Complete the factoring.
    a. $x^2 - 2x - 8 = (x + 2)(\quad)$ b. $x^2 + 12x + 35 = (x + 7)(\quad)$

13.   Factor completely: $x^2 + 3xy - 10y^2$

Factor as completely as possible. If unfactorable, indicate that the polynomial is prime.

14. $12x^2 + 17x + 6$
        a. $(12x + 2)(x + 3)$ b. $(3x - 2)(4x - 3)$
        c. $(3x + 2)(4x + 3)$ d. prime

15. $6x^2 - 5xt - 6t^2$
        a. prime   b. $(3x + 2t)(2x - 3t)$
        c. $(6x + 2t)(x - 3t)$ d. $(3x - 2t)(2x + 3t)$

16. $56 - 15x + x^2$
        a. $(x - 7)(x - 8)$ b. $(x + 7)(x + 8)$
        c. $(x + 7)(x - 8)$ d. $(x - 7)(x + 8)$

17. $15z^2 - 2z - 8$
        a. $(15z + 2)(z - 4)$ b. $(3z - 2)(5z + 4)$
        c. prime   d. $(3z + 2)(5z - 4)$

18. $12x^2 - 5xt - 3t^2$
        a. $(3x - t)(4x + 3t)$ b. $(3x + t)(4x - 3t)$
        c. $(12x + t)(x - 3t)$ d. prime

19.   Factor: $3x^2 + 25x + 8$

20.   Factor: $9h^2 - 24h + 16$

21.   Factor: $2a^6 + 8a^5 - 42a^4$

22.   Factor: $15x^2 - 19x + 6$

23.   Explain the error: $3k^3 - 12k^2 - 15k = (k - 5)(k + 1)$

24.   Factor: $7 + 8a + a^2$

Factor the following trinomials. There are NO primes.

1. $u^2 - 3uv - 54v^2$ 2. $u^2 - 5uv - 14v^2$

3. $s^2t^2 - 15st + 26$ 4. $x^2y^2 + 7xy + 12$

5. $-m^2 + 6m + 40$ 6. $x^2 + 3xy - 28y^2$

7. $x^2 - x - 42$ 8. $x^2 + 4xy - 192y^2$

9. $u^2 - 4uv - 45v^2$ 10. $u^2 - 2uv - 15v^2$

11. $12x^2 + 17x + 6$ 12. $15y^2 + 26y + 8$

13. $12z^2 - 7z - 12$ 14. $8z^2 + 6z - 9$

15. $24m^2n^2 - 79mn + 40$ 16. $28m^2 + 81mn + 56n^2$

17. $15x^2 + 26x + 8$

18. $6y^2 + 17y + 12$

19. $12z^2 + 7z - 12$

20. $20z^2 + 7z - 6$

21. $20 - 7y - 6y^2$

22. $343y^3 - 64z^3$

23. $6 + 2a - 3b - ab$

24. $-x^2 - x + 30$

25. $3x^2 - 5x - 12$

26. $y^4 - 64x^4$

27. $-x^2 + 5x - 6$

28. $1 + 64x^6$

29. $10x^2y^2 - 10y^3 + 15x^2y - 15y^2$

30. $y^2(x^3 + 8) - 1(x^3 + 8)$

LESSON 45 – Solving Quadratic Equations by Factoring

The general form of a quadratic equation is $Ax2 + Bx + C = 0$, where "A", "B", and "C" are real numbers and $A \neq 0$.

We must first set the quadratic equation equal to zero (0). Factor (if possible), then set each factor equal to zero (0).

Example 1: $a^2 + 5a + 4 = 0$

    Solution:

        Step 1:   Try to factor.

$$a^2 + 5a + 4 = 0$$
$$(a + 4)(a + 1) = 0$$

        Step 2:   Set each factor equal to zero (0).
        Step 3:   Solve each equation.

$$a + 4 = 0 \qquad a + 1 = 0$$
$$a = -4 \qquad\; a = -1$$

        Step 4:   Check both roots in the **original** equation.

$$
\begin{array}{ll}
a = -4 & a = -1 \\
a^2 + 5a + 4 = 0 & a^2 + 5a + 4 = 0 \\
(-4)2 + 5(-4) + 4 = 0 & (-1)2 + 5(-1) + 4 = 0 \\
16 - 20 + 4 = 0 & 1 - 5 + 4 = 0 \\
-4 + 4 = 0 & -4 + 4 = 0 \\
0 \overset{\checkmark}{=} 0 & 0 \overset{\checkmark}{=} 0
\end{array}
$$

    Solution: $\{-4, -1\}$

You'll notice that the answer (roots) are always put in braces { }. The order is immaterial. However, I prefer to list the smaller root first.

Example 2: $x^2 + 3x - 4 = 0$

    Solution:

        Step 1: $x^2 + 3x - 4 = 0$
                     $(x + 4)(x - 1) = 0$

        Steps 2 and 3: $x + 4 = 0 \qquad x - 1 = 0$
                            $x = -4 \qquad\; x = 1$

Step 4:   x = −4   x = 1

$$x^2 + 3x - 4 = 0$$ $$x^2 + 3x - 4 = 0$$
$$(-4)^2 + 3(-4) - 4 = 0$$ $$(1)^2 + 3(1) - 4 = 0$$
$$16 - 12 - 4 = 0$$ $$1 + 3 - 4 = 0$$
$$4 - 4 = 0$$ $$4 - 4 = 0$$
$$0 \overset{\checkmark}{=} 0$$ $$0 \overset{\checkmark}{=} 0$$

Solution: {−4, 1}

Example 3: $c^2 + 3c = 0$

Solution:

Step 1: $$c^2 + 3c = 0$$
            $$c(c + 3) = 0$$

Steps 2 and 3:   c = 0   c + 3 = 0
                                            c = −3

Step 4:   c = 0   c = −3

$$c^2 + 3c = 0$$ $$c^2 + 3c = 0$$
$$(0)^2 + 3(0) = 0$$ $$(-3)^2 + 3(-3) = 0$$
$$0 + 0 = 0$$ $$9 - 9 = 0$$
$$0 \overset{\checkmark}{=} 0$$ $$0 \overset{\checkmark}{=} 0$$

Solution: {−3, 0}

Example 4: $2a^2 - 3a - 2 = 0$

Solution:
Step 1: $$2a^2 - 3a - 2 = 0$$
            $$(2a + 1)(a - 2) = 0$$

Steps 2 and 3:   2a + 1 = 0   a − 2 = 0
                   $a = -\frac{1}{2}$ a = 2

Step 4: $a = -\frac{1}{2}$ $a = 2$

$2a^2 - 3a - 2 = 0$ $2a^2 - 3a - 2 = 0$

$2(-\frac{1}{2})^2 - 3(-\frac{1}{2}) - 2 = 0$ $2(2)^2 - 3(2) - 2 = 0$

$2(\frac{1}{4}) + \frac{3}{2} - 2 = 0$ $2(4) - 6 - 2 = 0$

$\frac{2}{4} + \frac{3}{2} - 2 = 0$ $8 - 6 - 2 = 0$

$\frac{1}{2} + \frac{3}{2} - 2 = 0$ $0 \overset{\checkmark}{=} 0$

$\frac{4}{2} - 2 = 0$

$2 - 2 = 0$

$0 \overset{\checkmark}{=} 0$

Solution: $\{-\frac{1}{2}, 2\}$

Example 5: $\dfrac{2}{b - 1} = \dfrac{b}{b + 2}$

Solution:

Cross-multiply: $2(b + 2) = b(b - 1)$
Expand: $2b + 4 = b^2 - b$
Set equal to 0: $0 = b^2 - 3b - 4$

Step 1: $0 = b^2 - 3b - 4$
            $0 = (b - 4)(b + 1)$

Steps 2 and 3: $b - 4 = 0$ $b + 1 = 0$
                 $b = 4$ $b = -1$

Step 4: $b = 4$ $b = -1$

$\dfrac{2}{b - 1} = \dfrac{b}{b + 2}$ $\dfrac{2}{b - 1} = \dfrac{b}{b + 2}$

$\dfrac{2}{4 - 1} = \dfrac{4}{4 + 2}$ $\dfrac{2}{-1 - 1} = \dfrac{-1}{-1 + 2}$

$\dfrac{2}{3} = \dfrac{4}{6}$ $\dfrac{2}{-2} = \dfrac{-1}{1}$

$\dfrac{2}{3} \overset{\checkmark}{=} \dfrac{2}{3}$ $-1 \overset{\checkmark}{=} -1$

Solution: $\{-1, 4\}$

Example 6: $4x^2 - 16 = 0$

    Solution:
        Step 1: $4x^2 - 16 = 0$
                              $4(x^2 - 4) = 0$
                              $4(x + 2)(x - 2) = 0$

        Steps 2 and 3: $4 \neq 0$ $x + 2 = 0$ $x - 2 = 0$
                            not a solution $x = -2$ $x = 2$

        Step 4: $x = -2$ $x = 2$

                        $4x^2 - 16 = 0$ $4x^2 - 16 = 0$
                        $4(-2)^2 - 16 = 0$ $4(2)^2 - 16 = 0$
                        $4(4) - 16 = 0$ $4(4) - 16 = 0$
                        $16 - 16 = 0$ $16 - 16 = 0$
                        $0 \overset{\checkmark}{=} 0$ $0 \overset{\checkmark}{=} 0$
    Solution: {-2, 2}

Let us try some practice problems. Solve and check each of the following.

1. $a^2 + 10 = 7a$ 2. $b^2 + 6b + 8 = 0$

3. $y^2 - y - 6 = 0$ 4. $x^2 - 8x = 0$

5. $2a^2 = 9a - 10$ 6. $m + 4 = 3m^2$

7. $a = \dfrac{24}{a - 2}$ 8. $2a^2 + a = 15$

9. $\dfrac{e - 3}{2} = \dfrac{8}{e + 3}$ 10. $4b^2 + 3b^2 = 63$

11. $2(a^2 + 1) = 5a$ 12. $2n^2 + 5n + 2 = 0$

13. $d^2 - 5d - 6 = 0$ 14. $5a^2 + 4 = 12a$

LESSON 46 – Irrational Numbers

An **IRRATIONAL NUMBER**, by definition, is an number which cannot be expressed as an integer or the quotient of two integers. The word **SURD** is given to any number that is irrational. Some examples of surds are $\sqrt{5}$, $\sqrt{17}$, $\sqrt[3]{20}$, $\sqrt[4]{7}$, $\sqrt{39}$ and π. Radicals that can be found exactly are rational. Some examples of rational numbers are $\sqrt{9}$, $\sqrt{16}$, $\sqrt{25}$), and many others. This understanding is necessary in order to comprehend Completing the Square and the Quadratic Formula, which are the topics of the next two lessons.

Here are some examples in simplifying a radical:

Example 1: $\sqrt{32}$

Solution: $32 = 2 \cdot 2 \cdot 2 \cdot 2 \cdot 2 = 4 \cdot 4 \cdot 2 = 16 \cdot 2$

$$\sqrt{32} = \sqrt{4 \cdot 4 \cdot 2} = \sqrt{4}\,\sqrt{4}\,\sqrt{2} = 2 \cdot 2 \cdot \sqrt{2} = 4\sqrt{2}$$

or

$$\sqrt{32} = \sqrt{16 \cdot 2} = \sqrt{16}\,\sqrt{2} = 4\sqrt{2}$$

Example 2: $\sqrt{48}$

Solution: $48 = 2 \cdot 2 \cdot 2 \cdot 2 \cdot 3 = 4 \cdot 4 \cdot 3 = 16 \cdot 3$

$$\sqrt{48} = \sqrt{4 \cdot 4 \cdot 3} = \sqrt{4}\,\sqrt{4}\,\sqrt{3} = 2 \cdot 2 \cdot \sqrt{3} = 4\sqrt{3}$$

or

$$\sqrt{48} = \sqrt{16 \cdot 3} = \sqrt{16}\,\sqrt{3} = 4\sqrt{3}$$

Example 3: $\sqrt{72}$

Solution: $72 = 2 \cdot 2 \cdot 2 \cdot 3 \cdot 3 = 4 \cdot 9 \cdot 2 = 36 \cdot 2$

$$\sqrt{72} = \sqrt{4 \cdot 9 \cdot 2} = \sqrt{4}\,\sqrt{9}\,\sqrt{2} = 2 \cdot 3 \cdot \sqrt{2} = 6\sqrt{2}$$

or

$$\sqrt{72} = \sqrt{36 \cdot 2} = \sqrt{36}\,\sqrt{2} = 6\sqrt{2}$$

Example 4: $\sqrt{112}$

   Solution: $112 = 2 \cdot 2 \cdot 2 \cdot 2 \cdot 7 = 4 \cdot 4 \cdot 7 = 16 \cdot 7$

   $\sqrt{112} = \sqrt{4 \cdot 4 \cdot 7} = \sqrt{4}\,\sqrt{4}\,\sqrt{7} = 2 \cdot 2 \cdot \sqrt{7} = 4\sqrt{7}$

   or

   $\sqrt{112} = \sqrt{16 \cdot 7} = \sqrt{16}\,\sqrt{7} = 4\sqrt{7}$

Example 5: $5\sqrt{8}$

   Solution: $8 = 2 \cdot 2 \cdot 2 = 4 \cdot 2$

   $5\sqrt{8} = 5\sqrt{4 \cdot 2} = 5\sqrt{4}\sqrt{2} = 5 \cdot 2 \cdot \sqrt{2} = 10\sqrt{2}$

Example 6: $2\sqrt{45}$

   Solution: $45 = 3 \cdot 3 \cdot 5 = 9 \cdot 5$

   $2\sqrt{45} = 2\sqrt{9 \cdot 5} = 2\sqrt{9}\sqrt{5} = 2 \cdot 3 \cdot \sqrt{5} = 6\sqrt{5}$

Now try these problems before we tackle the Quadratic Formula. Simplify the following.

1. $\sqrt{20}$ 2. $\sqrt{54}$ 3. $\sqrt{49}$

4. $3\sqrt{162}$ 5. $2\sqrt{40}$ 6. $\sqrt{200}$

7. $\frac{1}{3}\sqrt{50}$ 8. $\sqrt{150}$ 9. $5\sqrt{16}$

10. $2\sqrt{63}$ 11. $\sqrt{96}$ 12. $3\sqrt{80}$

13. $\frac{1}{2}\sqrt{48}$ 14. $2\sqrt{300}$ 15. $\sqrt{144}$

16. $\sqrt{98}$ 17. $4\sqrt{28}$ 18. $\sqrt{288}$

LESSON 47 – Solving Quadratic Equations by Completing the Square

Lesson 45 showed how to solve a quadratic equation, $Ax^2 + Bx + C = 0$, where A, B, and C are real numbers and $A \neq 0$, by factoring. However, not every quadratic equation can be factored by using integers.

Consider this equation: $x^2 - 4x + 1 = 0$. Since it cannot factor easily, we will use a method know as **Completing the Square**.

The step for Completing the Square are:

Step 1:   Collect all variables on the left side of the equation and all constants on the right side of the equation.

Step 2:   Divide B by 2. Square the result and add to both sides of the equation. Simplify the equation

Step 3:   Since the left side of the equation is now a perfect square, we can factor it.

Step 4:   Find the square root of both sides.

Step 5:   Solve for "x".

Example 1: $x^2 - 4x + 1 = 0$

Solution: B = –4

Step 1: $x^2 - 4x = -1$

Step 2:   B = –4
$$\frac{B}{2} = \frac{-4}{2} = -2$$
$$(-2)^2 = 4$$

$x^2 - 4x + 4 = -1 + 4$
$x^2 - 4x + 4 = 3$

Adding 4 to $x^2 - 4x$ is an example of Completing the Square. The trinomial $x^2 - 4x + 4$ is the square of $x - 2$.

Step 3: $(x - 2)^2 = 3$

Step 4: $\sqrt{(x-2)^2} = \pm\sqrt{3}$
$x - 2 = \pm\sqrt{3}$

Step 5: $x - 2 = \pm\sqrt{3}$

$x = 2 \pm \sqrt{3}$

Solution: The original equation, $x^2 - 4x + 1 = 0$, has two solutions;
$2 + \sqrt{3}$ and $2 - \sqrt{3}$.

Example 2: $x^2 - 6x + 2 = 0$

Solution: $B = -6$

Step 1: $x^2 - 6x = -2$

Step 2: $x^2 - 6x + 9 = -2 + 9$
$x^2 - 6x + 9 = 7$

Step 3: $(x - 3)^2 = 7$

Step 4: $\sqrt{(x - 3)^2} = \pm\sqrt{7}$
$x - 3 = \pm\sqrt{7}$

Step 5: $x = 3 \pm \sqrt{7}$

Solution: $x = 3 + \sqrt{7}$ and $x = 3 - \sqrt{7}$

Check:
$(3 + \sqrt{7})^2 - 6(3 + \sqrt{7}) + 2 = 0$
$(3 + \sqrt{7})(3 + \sqrt{7}) - 18 - 6\sqrt{7} + 2 = 0$
$9 + 3\sqrt{7} + 3\sqrt{7} + 7 - 18 - 6\sqrt{7} + 2 = 0$
$9 + 6\sqrt{7} + 7 - 18 - 6\sqrt{7} + 2 = 0$
$(9 + 7 - 18 + 2) + (6\sqrt{7} - 6\sqrt{7}) = 0$
$16 - 18 + 2 + 0 = 0$
$-2 + 2 = 0$
$0 \overset{\checkmark}{=} 0$

$(3 - \sqrt{7})^2 - 6(3 - \sqrt{7}) + 2 = 0$
$(3 - \sqrt{7})(3 - \sqrt{7}) - 18 + 6\sqrt{7} + 2 = 0$
$9 - 3\sqrt{7} - 3\sqrt{7} + 7 - 18 + 6\sqrt{7} + 2 = 0$
$9 - 6\sqrt{7} + 7 - 18 - 6\sqrt{7} + 2 = 0$
$(9 + 7 - 18 + 2) + (-6\sqrt{7} + 6\sqrt{7}) = 0$
$16 - 18 + 2 + 0 = 0$
$-2 + 2 = 0$
$0 \overset{\checkmark}{=} 0$

Example 3: $x^2 - 8x = 2$

Solution: $B = -8$

Step 1: $x^2 - 8x = 2$

Step 2: $x^2 - 8x + 16 = 2 + 16$

Step 3: $(x - 4)^2 = 18$

Step 4: $\sqrt{(x - 4)^2} = \pm\sqrt{18}$
$x - 4 = \pm\sqrt{18}$
$x - 4 = \pm 3\sqrt{2}$

Step 5: $x = 4 \pm 3\sqrt{2}$

Solution: $4 + 3\sqrt{2}$ and $4 - \sqrt{2}$

Check:
$(4 + 3\sqrt{2})^2 - 8(4 + 3\sqrt{2}) = 2$
$(4 + 3\sqrt{2})(4 + 3\sqrt{2}) - 32 - 24\sqrt{2} = 2$
$16 + 12\sqrt{2} + 12\sqrt{2} + 18 - 32 - 24\sqrt{2} = 2$
$16 + 24\sqrt{2} + 18 - 32 - 24\sqrt{2} = 2$
$(16 + 18 - 32) + (24\sqrt{2} - 24\sqrt{2}) = 2$
$34 - 32 + 0 = 2$
$2 \overset{\checkmark}{=} 2$

$(4 - 3\sqrt{2})^2 - 8(4 - 3\sqrt{2}) = 2$
$(4 - 3\sqrt{2})(4 - 3\sqrt{2}) - 32 + 24\sqrt{2} = 2$
$16 - 12\sqrt{2} - 12\sqrt{2} + 18 - 32 + 24\sqrt{2} = 2$
$16 - 24\sqrt{2} + 18 - 32 + 24\sqrt{2} = 2$
$(16 + 18 - 32) + (-24\sqrt{2} + 24\sqrt{2}) = 2$
$34 - 32 + 0 = 2$
$2 \overset{\checkmark}{=} 2$

Example 4: Solve $x^2 + 10x - 9 = 0$.

Solution: B = 10

Step 1: $x^2 + 10x = 9$

Step 2: $x^2 + 10x + 25 = 9 + 25$
           $x^2 + 10x + 25 = 34$

Step 3: $(x + 5)^2 = 34$

Step 4: $\sqrt{(x + 5)^2} = \pm\sqrt{34}$
           $x + 5 = \pm\sqrt{34}$

Step 5: $x = -5 \pm \sqrt{34}$

Solution: $-5 + \sqrt{34}$ and $-5 - \sqrt{34}$

Practice problems. Solve each equation by completing the square. Express all irrational roots in simplest form.

1. $x^2 + 2x - 5 = 0$　　　　　2. $x^2 + 6x - 5 = 0$

3. $x^2 = 3 - x$　　　　　　　4. $x^2 - 5x + 1 = 0$

5. $x^2 + 4x = -2$　　　　　　6. $x^2 + 4x - 3 = 0$

7. $2x^2 - 3x - 2 = 0$　　　　8. $y^2 + 4y = 21$
    Hint: Divide all terms of the equation
    by the coefficient of x2.

9. $x^2 + 2x - 48 = 0$　　　　10. $x^2 - 2x - 5 = 0$

11. $2x^2 - 10x = 1$　　　　　12. $x^2 - 6x = 16$

LESSON 48 – Solving Quadratic Equations Using the Quadratic Formula

The coefficients of the general formula, $ax^2 + bx + c = 0$, are used in the Quadratic Formula:

$$x = \frac{-b \pm \sqrt{b^2 - 4ac}}{2a}$$

where "a", "b", and "c" are obtained from the equation.

Steps for solving a quadratic equation:

Step 1: Simplify the equation and transpose all terms so that they are on the same side of the equation and equal to 0.

Step 2: Select the values of "a", "b" and "c" as shown in the Quadratic Formula

Step 3: Substitute the values for "a", "b" and "c" into the Quadratic Formula.

Step 4: Simplify the Quadratic Formula and solve for "x".

Step 5: Check the roots in the **original** quadratic equation.

Let us look closely at the first example as we proceed step-by-step.

Example 1: $x^2 + 3x - 40 = 0$

Solution:

Step 1: Already equal to 0.

Step 2: $a = 1, b = 3, c = -40$

Step 3: $x = \dfrac{-b \pm \sqrt{b^2 - 4ac}}{2a} = \dfrac{-3 \pm \sqrt{(3)^2 - 4(1)(-40)}}{2(1)}$

Step 4: $\dfrac{-3 \pm \sqrt{(3)^2 - 4(1)(-40)}}{2(1)} = \dfrac{-3 \pm \sqrt{9 + 160}}{2} =$

$\dfrac{-3 \pm \sqrt{169}}{2} = \dfrac{-3 \pm 13}{2}$

$x = \dfrac{-3 + 13}{2} = \dfrac{10}{2} = 5 \qquad x = \dfrac{-3 - 13}{2} = \dfrac{-16}{2} = -8$

Step 5:   x = 5   x = –8
            $x^2 + 3x - 40 = 0$ $x^2 + 3x - 40 = 0$
            $(5)^2 + 3(5) - 40 = 0$ $(-8)^2 + 3(-8) - 40 = 0$
            $25 + 15 - 40 = 0$ $64 - 24 - 40 = 0$
            $40 - 40 = 0$ $40 - 40 = 0$
            $0 \overset{\checkmark}{=} 0$ $0 \overset{\checkmark}{=} 0$

Solution: {–8, 5}

Example 2: $x^2 + 4x = 21$

        Solution:
            Step 1: $x^2 + 4x - 21 = 0$

            Step 2:   a = 1, b = 4, c = –21

            Step 3: $x = \dfrac{-b \pm \sqrt{b^2 - 4ac}}{2a} = \dfrac{-4 \pm \sqrt{(4)^2 - 4(1)(-21)}}{2(1)}$

            Step 4: $x = \dfrac{-4 \pm \sqrt{(4)^2 - 4(1)(-21)}}{2(1)} = \dfrac{-4 \pm \sqrt{16 + 84}}{2} =$

                        $\dfrac{-4 \pm \sqrt{100}}{2} = \dfrac{-4 \pm 10}{2}$

                        $x = \dfrac{-4 + 10}{2} = \dfrac{6}{2} = 3$ $x = \dfrac{-4 - 10}{2} = \dfrac{-14}{2} = -7$

            Step 5:   x = 3   x = –7
                        $x^2 + 4x = 21$ $x^2 + 4x = 21$
                        $(3)^2 + 4(3) = 21$ $(-7)^2 + 4(-7) = 21$
                        $9 + 12 = 21$ $49 - 28 = 21$
                        $21 \overset{\checkmark}{=} 21$ $21 \overset{\checkmark}{=} 21$

        Solution: {–7, 3}

Example 3: $2x^2 - 3x - 2 = 0$

Solution:

Step 1:   Already equal to 0.

Step 2: $a = 2, b = -3, c = -2$

Step 3: $x = \dfrac{-b \pm \sqrt{b^2 - 4ac}}{2a} = \dfrac{3 \pm \sqrt{(-3)^2 - 4(2)(-2)}}{2(2)}$

Step 4: $x = \dfrac{3 \pm \sqrt{(-3)^2 - 4(2)(-2)}}{2(2)} = \dfrac{3 \pm \sqrt{9 + 16}}{4} = \dfrac{3 \pm \sqrt{25}}{4} = \dfrac{3 \pm 5}{4}$

$x = \dfrac{3 + 5}{4} = \dfrac{8}{4} = 2$ $\qquad$ $x = \dfrac{3 - 5}{4} = \dfrac{-2}{4} = -\dfrac{1}{2}$

Step 5: $x = 2$ $\qquad\qquad\qquad\qquad$ $x = -\dfrac{1}{2}$

$2x^2 - 3x = 0$ $\qquad\qquad\qquad$ $2x^2 - 3x = 0$

$2(2)^2 - 3(2) - 2 = 0$ $\qquad\quad$ $2(-\tfrac{1}{2})^2 - 3(-\tfrac{1}{2}) - 2 = 0$

$2(4) - 6 - 2 = 0$ $\qquad\qquad$ $2(\tfrac{1}{4}) + \tfrac{3}{2} - 2 = 0$

$8 - 6 - 2 = 0$ $\qquad\qquad\quad$ $\tfrac{1}{2} + \tfrac{3}{2} - 2 = 0$

$2 - 2 = 0$ $\qquad\qquad\qquad$ $\tfrac{4}{2} - 2 = 0$

$0 \overset{\checkmark}{=} 0$ $\qquad\qquad\qquad\;$ $2 - 2 = 0$

$\qquad\qquad\qquad\qquad\qquad\quad$ $0 \overset{\checkmark}{=} 0$

Solution: $\{-\tfrac{1}{2}, 2\}$

Example 4: $2x^2 + 4x = 3$

Solution:

Step 1: $2x^2 + 4x - 3 = 0$

Step 2: $a = 2, b = 4, c = -3$

Step 3: $x = \dfrac{-b \pm \sqrt{b^2 - 4ac}}{2a} = \dfrac{-4 \pm \sqrt{(4)^2 - 4(2)(-3)}}{2(2)}$

Step 4: $x = \dfrac{-4 \pm \sqrt{(4)^2 - 4(2)(-3)}}{2(2)} = \dfrac{-4 \pm \sqrt{16 + 24}}{4} =$

$\dfrac{-4 \pm \sqrt{40}}{4} = \dfrac{-4 \pm 2\sqrt{10}}{4}$

$$x = \frac{-4 + 2\sqrt{10}}{4} = \frac{-4}{4} + \frac{2\sqrt{10}}{4} = -1 + \frac{\sqrt{10}}{2}$$

NOTE: The denominator (4) is used with both terms.

$$x = \frac{-4 - 2\sqrt{10}}{4} = \frac{-4}{4} - \frac{2\sqrt{10}}{4} = -1 - \frac{\sqrt{10}}{2}$$

Step 5:

$x = -1 + \frac{\sqrt{10}}{2}$

$2x^2 + 4x = 3$

$2(-1 + \frac{\sqrt{10}}{2})^2{}^* + 4(-1 + \frac{\sqrt{10}}{2}) = 3$

$2(\frac{7}{2} - \sqrt{10}) - 4 + 2\sqrt{10} = 3$

$7 - 2\sqrt{10} - 4 + 2\sqrt{10} = 3$

$(7 - 4) + (-2\sqrt{10} + 2\sqrt{10}) = 3$

$3 + 0 = 3$

$3 \overset{\checkmark}{=} 3$

$x = -1 - \frac{\sqrt{10}}{2}$

$2x^2 + 4x = 3$

$2(-1 - \frac{\sqrt{10}}{2})^2{}^{**} - 3(-1 - \frac{\sqrt{10}}{2}) = 0$

$2(\frac{7}{2} + \sqrt{10}) - 4 - 2\sqrt{10} = 3$

$7 + 2\sqrt{10} - 4 - 2\sqrt{10} = 3$

$(7 - 4) + (2\sqrt{10} - 2\sqrt{10}) = 3$

$3 + 0 = 3$

$3 \overset{\checkmark}{=} 3$

Solution: $\{-1 \pm \frac{\sqrt{10}}{2}\}$

*

$(-1 + \frac{\sqrt{10}}{2})^2 = (-1 + \frac{\sqrt{10}}{2})(-1 + \frac{\sqrt{10}}{2})$

Using FOIL:

F: $-1 \cdot -1 = 1$

O: $-1 \cdot \frac{\sqrt{10}}{2} = -\frac{\sqrt{10}}{2}$

I: $-1 \cdot \frac{\sqrt{10}}{2} = -\frac{\sqrt{10}}{2}$

L: $\frac{\sqrt{10}}{2} \cdot \frac{\sqrt{10}}{2} = \frac{\sqrt{100}}{4} = \frac{10}{4} = \frac{5}{2}$

$(-1 + \frac{\sqrt{10}}{2})^2 = 1 - \frac{\sqrt{10}}{2} - \frac{\sqrt{10}}{2} + \frac{5}{2} =$

$1 - \frac{2\sqrt{10}}{2} + \frac{5}{2} = 1 - \sqrt{10} + \frac{5}{2} =$

$(1 + \frac{5}{2}) - \sqrt{10} = \frac{7}{2} - \sqrt{10}$

$(-1 + \frac{\sqrt{10}}{2})^2 = \frac{7}{2} - \sqrt{10}$

**

$(-1 - \frac{\sqrt{10}}{2})^2 = (-1 - \frac{\sqrt{10}}{2})(-1 - \frac{\sqrt{10}}{2})$

Using FOIL:

F: $-1 \cdot -1 = 1$

O: $-1 \cdot -\frac{\sqrt{10}}{2} = \frac{\sqrt{10}}{2}$

I: $-1 \cdot -\frac{\sqrt{10}}{2} = \frac{\sqrt{10}}{2}$

L: $-\frac{\sqrt{10}}{2} \cdot -\frac{\sqrt{10}}{2} = \frac{\sqrt{100}}{4} = \frac{10}{4} = \frac{5}{2}$

$(-1 - \frac{\sqrt{10}}{2})^2 = 1 + \frac{\sqrt{10}}{2} + \frac{\sqrt{10}}{2} + \frac{5}{2} =$

$1 + \frac{2\sqrt{10}}{2} + \frac{5}{2} = 1 + \sqrt{10} + \frac{5}{2} =$

$(1 + \frac{5}{2}) + \sqrt{10} = \frac{7}{2} + \sqrt{10}$

$(-1 - \frac{\sqrt{10}}{2})^2 = \frac{7}{2} + \sqrt{10}$

Try these for practice. Use the Quadratic Formula and leave answers in square root form.

1. $x^2 - 10x - 15 = 0$ 2. $3x^2 = 5x + 4$

3. $3x^2 - 2x - 6 = 0$ 4. $x^2 + 2x - 4 = 0$

5. $x^2 - 3x - 3 = 0$ 6. $x^2 + 5x - 6 = 0$

7. $x^2 + 6x - 4 = 0$ 8. $x^2 - 2x - 5 = 0$

9. $x^2 = 4x + 12$ 10. $x^2 = 14 - 5x$

11. $2x^2 + 7x + 1 = 0$ 12. $3x^2 + 5x + 1 = 0$

13. $2x^2 - 5x + 1 = 0$ 14. $x^2 + 9x - 12 = 0$

15. $x^2 = 20x + 10$ 16. $2x^2 = 8x - 1$

17. $x^2 + 4x - 20 = -4$ 18. $2x^2 = 2x + 3$

19. $2x^2 - 10x = 1$ 20. $x^2 = 2 - 5x$

LESSON 49 – A Review of Lessons 45-48

Solve the following quadratic equations using the Quadratic Formula or factoring.

1. $x^2 + 16 = 8x$

2. $3x^2 + 5x = -2$

3. $2 + 4x = 5x^2$

4. $2(x^2 - 3) = x$

5. $y^2 + 4y - 2 = 0$

6. $2x^2 - 10x = 9$

7. $x^2 - 8x + 12 = 0$

8. $2x^2 + x = 15$

9. $5x^2 + 3x = 2$

10. $x^2 - 5x - 4 = 0$

11. $3x^2 + 2x - 1 = 0$

12. $c^2 - 6c = 16$

13. $2x^2 - 4x - 1 = 0$

14. $x^2 + 2x - 48 = 0$

15. $x^2 - 2x = 4$

16. $4x^2 - 3x - 8 = 0$

LESSON 50 – Systems of Equations Solved by Elimination by Either Addition or Subtraction

The procedure is as follows:

Step 1: Simplify both equations by removing parentheses and/or clearing fractions, if need be.

Step 2: Put the unknown quantities on the left side of the equal signs and the constants to the other side and then combine any like terms.

Step 3A: Select a variable to eliminate in both equations. If the constant of the variable is the same for both equations and the signs are opposite, add the two equations. If not, subtract.

Step 3B: If the two constants are unlike, multiply each equation by the other coefficient of the variable. Then proceed to add or subtract, depending upon the signs.

Step 4: Once one of the two variables is eliminated, the result is a simple equation which you can solve.

Step 5: Substitute that value in one of the **original** equations and solve.

Step 6: The result should give you solutions for BOTH variables which need to be checked for correctness.

Let us review several examples.

Example 1: $\begin{cases} x + y = 10 & \text{(A)} \\ x - y = 4 & \text{(B)} \end{cases}$

Solution:

Step 1: Since there are no parentheses or fractions, proceed to step 2.

Step 2: In both equations, the variables are on one side and the constants on the other, so we go the step 3.

Step 3: Select a variable to eliminate. If the coefficients are identical and have opposite signs, choose it. In the example, choose "y".

$$
\begin{array}{rcrcr}
x & + & y & = & 10 \\
x & - & y & = & 4 \\
\hline
2x & + & 0 & = & 14
\end{array}
$$
Add both equations together.

136

Step 4: Solve for "x".

$$2x = 14$$
$$x = 7$$

Step 5: Substitute the value of "x" in either **original** equation, and solve for "y".

$x + y = 10$	or	$x - y = 4$
$7 + y = 10$		$7 - y = 4$
$y = 3$		$y = 3$

Step 6: Check it!

Equation (A)	Equation (B)
$x + y = 10$	$x - y = 4$
$7 + 3 = 10$	$7 - 3 = 4$
$10 \overset{\checkmark}{=} 10$	$4 \overset{\checkmark}{=} 4$

Solution: $x = 7$ and $y = 3$

Example 2: $\begin{cases} 5x + y = 4 & \text{(A)} \\ 3x - 2y = -8 & \text{(B)} \end{cases}$

Solution:

Step 1: No parentheses or fractions. Proceed to step 2.

Step 2: Variables on the left, constants on the right. Proceed to step 3.

Step 3: Select a variable to eliminate, "y". Why "y"? The "y's" have opposite signs and you only need to multiply equation (A) by 2.

$$2 \cdot (\ 5x \ + \ y \ = \ 4 \ ) \ \rightarrow \ \begin{array}{r} 10x + 2y = 8 \\ 3x - 2y = -8 \\ \hline 13x \qquad = 0 \end{array}$$

$$3x - 2y = -8$$

Step 4: Solve for "x".

$$13x = 0$$
$$x = 0$$

Step 5: Substitute x = 0 in either of the **original** equations.

Equation (A)   or   Equation (B)
5x + y = 4   3x − 2y = −8
5(0) + y = 4   3(0) − 2y = −8
y = 4   −2y = −8
                                    y = 4

Step 6: Check it!

Equation (A)   or   Equation (B)
5(0) + 4 = 4   3(0) − 2(4) = −8
0 + 4 = 4   0 − 8 = −8
4 $\overset{\checkmark}{=}$ 4   −8 $\overset{\checkmark}{=}$ −8

Solution: x = 0 and y = 4

Example 3: $\begin{cases} m = 11 + n & \text{(A)} \\ 3m = 3 - 2n & \text{(B)} \end{cases}$

Solution:

Step 1:   No parentheses or fractions. Proceed to step 2.

Step 2:   Need to move the "n" in both equations to the left side. When doing this, remember that the sign of the variable changes when crossing the equal sign.

m − n = 11   (A)
3m + 2n = 3   (B)

Step 3:   Select a variable, "n" to eliminate. See example 2 Step 3 as to why

$$2 \cdot (\quad m - n = 11 \quad) \quad \rightarrow \quad \begin{array}{rcrcr} 2m & - & 2n & = & 22 \\ 3m & + & 2n & = & 3 \\ \hline 5m & & & = & 25 \end{array}$$

Step 4:   Solve for "m".

5m = 25
m = 5

138

Step 5:   Substitute m = 5 into either **original** equation.

Equation (A)   or   Equation (B)
m = 11 + n   3m = 3 − 2n
5 = 11 + n   3(5) = 3 − 2n
−6 = n   15 = 3 − 2n
                                  12 = −2n
                                  −6 = n

Step 6:   Check it!

Equation (A)   or   Equation (B)
m = 11 + n   3m = 3 − 2n
5 = 11 + (−6)   3(5) = 3 − 2(−6)
5 $\overset{\checkmark}{=}$ 5   15 = 3 + 12
                                  15 $\overset{\checkmark}{=}$ 15

Solution: m = 5 and n = −6

Example 4: $\begin{cases} \dfrac{3x + 8}{5} = \dfrac{3y - 1}{2} \quad \text{(A)} \\ \dfrac{x + y}{2} = 3 + \dfrac{x - y}{2} \quad \text{(B)} \end{cases}$

Solution:
Step 1:   ALERT!!! WE HAVE FRACTIONS!!

In equation (A), cross multiply.

(3x + 8)(2) = (5)(3y − 1)
6x + 16 = 15y − 5

In equation (B), remove the fractions by multiplying **_all_** three terms by 2.

$2\left(\dfrac{x + y}{2}\right) = 2(3) + 2\left(\dfrac{x - y}{2}\right)$
x + y = 6 + x − y

Step 2:   Variables to the left, constants to the right. Combine like terms.

6x − 15y = −21
2y = 6   (the "x" canceled out)

139

Step 3:   Solve for "y".

$$2y = 6$$
$$y = 3$$

Step 4:   Substitute $y = 3$ into the original equation (A).

$$\frac{3x + 8}{5} = \frac{3(3) - 1}{2}$$
$$\frac{3x + 8}{5} = \frac{9 - 1}{2}$$
$$\frac{3x + 8}{5} = \frac{8}{2}$$

Step 5:   Solve for "x".

$$2(3x + 8) = 40 \qquad \text{cross multiply}$$
$$6x + 16 = 40$$
$$6x = 24$$
$$x = 4$$

Step 6:   Check it!

Equation (A)   Equation (B)

$$\frac{3x + 8}{5} = \frac{3y - 1}{2} \qquad\qquad \frac{x + y}{2} = 3 + \frac{x - y}{2}$$
$$\frac{3(4) + 8}{5} = \frac{3(3) - 1}{2} \qquad\qquad \frac{4 + 3}{2} = 3 + \frac{4 - 3}{2}$$
$$\frac{3(4) + 8}{5} = \frac{9 - 1}{2} \qquad\qquad \frac{7}{2} = 3 + \frac{1}{2}$$
$$\frac{3(4) + 8}{5} = \frac{8}{2} \qquad\qquad \frac{7}{2} = 3\tfrac{1}{2}$$
$$\frac{12 + 8}{5} = \frac{8}{2} \qquad\qquad 3\tfrac{1}{2} \overset{\checkmark}{=} 3\tfrac{1}{2}$$
$$\frac{20}{5} = \frac{8}{2}$$
$$4 \overset{\checkmark}{=} 4$$

Solution: $x = 4$ and $y = 3$

Now you can try some. Use the examples to help you succeed.

1. $6x + 10y = 7$
$15x - 4y = 3$

2. $2a - b = 8$
$a + 2b = 9$

3. $x - y = -1$
$3x - 2y = 3$

4. $5m - 2n = 3$
$2m - n = 0$

5. $3a + 7b = -4$
$2a + 5b = -3$

6. $2x + 3y = 17$
$3x + 5y = 27$

7. $3a = b + 1$
$5a = 3b + 7$

8. $\frac{1}{2}a + \frac{1}{3}b = 8$
$\frac{3}{2}a - \frac{4}{3}b = -4$

9. $5e = 4f$
$\frac{e}{2} + 2f = 12$

10. $a + b = 800$
$0.02a = 0.03b + 1$

11. $\frac{a}{3} + \frac{b}{2} = \frac{5}{6}$
$\frac{a}{2} + \frac{b}{3} = \frac{5}{6}$

12. $a + 2b = 3$
$3a - 2b = 8$

LESSON 51 – Systems of Equations Solved by Substitution

The procedure for solving by substitution is as follows:

Example 1: $\begin{cases} 6a + x = 12 & \text{(A)} \\ 2a + x = 8 & \text{(B)} \end{cases}$

Solution:

Step 1: Using one of the equations, solve for one of the unknown variables in terms of the other variable.

Equation (A)	Equation (B)
$6a + x = 12$	$2a + x = 8$
$x = -6a + 12$	$x = -2a + 8$

Step 2: Substitute this into the other equation.

Equation (B)	Equation (A)
$2a + x = 8$	$6a + x = 12$
$2a + (-6a + 12) = 8$	$6a + (-2a + 8) = 12$

Step 3: Solve for "a".

Equation (B)	Equation (A)
$-4a + 12 = 8$	$4a + 8 = 12$
$-4a = -4$	$4a = 4$
$a = 1$	$a = 1$

Step 4: Substitute the known value (a = 1) into the other equation.

Equation (A)	Equation (B)
$6a + x = 12$	$2a + x = 8$
$6(1) + x = 12$	$2(1) + x = 8$
$6 + x = 12$	$2 + x = 8$
$x = 6$	$x = 6$

Step 5. Check it!

Equation (A)	Equation (B)
$6a + x = 12$	$2a + x = 8$
$6(1) + 6 = 12$	$2(1) + 6 = 8$
$6 + 6 = 12$	$2 + 6 = 8$
$12 \overset{\checkmark}{=} 12$	$8 \overset{\checkmark}{=} 8$

Solution: a = 1 and x = 6

Example 2:
$$\begin{cases} 2a - 5b = 3 & \text{(A)} \\ 3b + a = 7 & \text{(B)} \end{cases}$$

Solution:

Step 1: Solve for "a" in equation (B).

$$3b + a = 7$$
$$a = -3b + 7$$

Step 2: Substitute for "a" in equation (A).

$$2a - 5b = 3$$
$$2(-3b + 7) - 5b = 3$$

Step 3: Solve for "b".

$$-6b + 14 - 5b = 3$$
$$-11b + 14 = 3$$
$$-11b = -11$$
$$b = 1$$

Step 4: Substitute b = 1 into equation (B).

$$3b + a = 7$$
$$3(1) + a = 7$$
$$3 + a = 7$$
$$a = 4$$

Step 5: Check it!

Equation (A)   Equation (B)
$$2a - 5b = 3 \qquad\qquad 3b + a = 7$$
$$2(4) - 5(1) = 3 \qquad\qquad 3(1) + 4 = 7$$
$$8 - 5 = 3 \qquad\qquad 3 + 4 = 7$$
$$3 \overset{\checkmark}{=} 3 \qquad\qquad 7 \overset{\checkmark}{=} 7$$

Solution: a = 4 and b = 1

Example 3: $\begin{cases} 2x - 5y = -4 & \text{(A)} \\ x + y = 5 & \text{(B)} \end{cases}$

Solution:

Step 1:   Solve for "y" in equation (B).

$$x + y = 5$$
$$y = -x + 5$$

Step 2:   Substitute for y in equation (A).

$$2x - 5y = -4$$
$$2x - 5(-x + 5) = -4$$

Step 3:   Solve for "x".

$$2x + 5x - 25 = -4$$
$$7x - 25 = -4$$
$$7x = 21$$
$$x = 3$$

Step 4:   Substitute x = 3 into equation (B).

$$x + y = 5$$
$$3 + y = 5$$
$$y = 2$$

Step 5:   Check your work!

Equation (A)	Equation (B)
$2x - 5y = -4$	$x + y = 5$
$2(3) - 5(2) = -4$	$3 + 2 = 5$
$6 - 10 = -4$	$5 \overset{\checkmark}{=} 5$
$-4 \overset{\checkmark}{=} -4$	

Solution: x = 3 and y = 2

Example 4: $\begin{cases} m + n = 7 & \text{(A)} \\ 2m - n = -1 & \text{(B)} \end{cases}$

Solution:

Step 1:   Solve for "n" in equation (A).

$$m + n = 7$$
$$n = -m + 7$$

Step 2:   Substitute for "n" in equation (B).

$$2m - n = -1$$
$$2m - (-m + 7) = -1$$

Step 3:   Solve for "m".

$$2m + m - 7 = -1$$
$$3m - 7 = -1$$
$$3m = 6$$
$$m = 2$$

Step 4:   Substitute m = 2 into equation (A).

$$m + n = 7$$
$$2 + n = 7$$
$$n = 5$$

Step 5:   Check your work!

Equation (A)	Equation (B)
$m + n = 7$	$2m - n = -1$
$2 + 5 = 7$	$2(2) - 5 = -1$
$7 \overset{\checkmark}{=} 7$	$4 - 5 = -1$
	$-1 \overset{\checkmark}{=} -1$

Solution: m = 2 and n = 5

Solve each set of equations by substitution. Check your answers.

1. $a = b + 1$
   $5a - 3b = 9$

2. $x = -y$
   $4x + 3y = 3$

3. $2c - d = 1$
   $d - c = 2$

4. $4e + 5f = 7$
   $3e + 4f = 6$

5. $8a + \frac{1}{4}b = 7$
   $4a + b = 0$

6. $x = 2z + 1$
   $x + z = -2$

7. $b = 3a - 1$
   $9a + 2b = 3$

8. $p = 3r$
   $p + r = 8$

9. $3m = 4n$
   $4m - 5n = 2$

10. $5x + 2 = -y$
    $x - 5 = -2y$

11. $x - y = 3$
    $2x - 3y = 1$

12. $4a - 5b = 7$
    $2a = 9 - 3b$

LESSON 52 – Systems of Equations Solved by Graphing

Of the many ways to solve systems of equations, graphing is the least used. The reason being, not everyone has access to a graphing calculator. Furthermore, finding graph paper or creating your own does present accuracy problems. Let us try to solve systems of equations by graphing the following examples.

Example 1: $\begin{cases} .x + 2y = 8 & \text{(A)} \\ y - x = 4 & \text{(B)} \end{cases}$

Solution:

Step 1: Put both equations in the form $y = mx + b$. Refer to Lesson 24 if you need to review.

Equation (A)

$x + 2y = 8$

$2y = -x + 8$

$y = -\frac{1}{2}x + 4$

Equation (B)

$y - x = 4$

$y = x + 4$

Step 2: Create a grid and graph both equations on it.

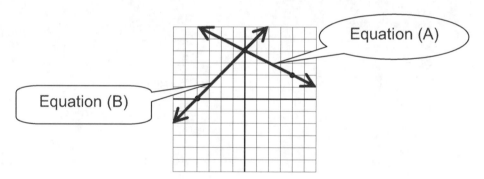

Step 3: Locate where both lines intersect: $x = 0$ and $y = 4$ (0, 4). This will be the solution set for both equations.

Step 4: Check your solution in **_both_** equations.

Equation (A)

$x + 2y = 8$

$0 + 2(4) = 8$

$8 \overset{\checkmark}{=} 8$

Equation (B)

$y - x = 4$

$4 - 0 = 4$

$4 \overset{\checkmark}{=} 4$

Solution: $x = 0$ and $y = 4$

We have solved this system of equations by graphing.

Example 2: $\begin{cases} y + x = 4 & \text{(A)} \\ 2y - x = -1 & \text{(B)} \end{cases}$

Solution:

Step 1:   Put both equations in the form y = mx + b.

Equation (A)   Equation (B)
    y + x = 4   2y − x = −1
    y = −x + 4   2y = x − 1
                                     $y = \frac{1}{2}x - \frac{1}{2}$

Step 2:   Create a grid and graph both equations on it.

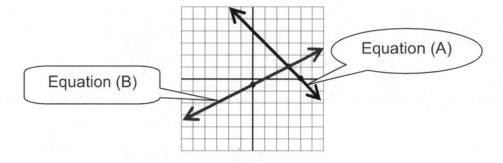

Step 3:   Locate the intersection: x = 3, y = 1 (3, 1)

Step 4:   Check it!

Equation (A)   Equation (B)
    y + x = 4   2y − x = −1
    1 + 3 = 4   2(1) − 3 = −1
    $4 \overset{\checkmark}{=} 4$ 2 − 3 = −1
                                     $-1 \overset{\checkmark}{=} -1$

Solution: x = 3 and y = 1

Example 3: $\begin{cases} 2x - 3y = 12 & \text{(A)} \\ y = -\frac{1}{3}x - 5 & \text{(B)} \end{cases}$

Solution:

Step 1:   Put both equations in the form y = mx + b.

Equation (A)   Equation (B)
2x − 3y = 12 $y = -\frac{1}{3}x - 5$
−3y = −2x + 12   (already in "y" form)
$y = \frac{2}{3}x - 4$

Step 2:   Create a grid and graph both equations.

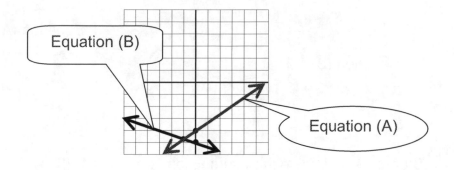

Step 3:   Locate the solution: x = −1 and $y = -\frac{14}{3} = -4\frac{2}{3}$ $(-1, -4\frac{2}{3})$

Step 4:   Check it!

Equation (A)   Equation (B)
2x − 3y = 12 $y = -\frac{1}{3}x - 5$
$2(-1) - 3(-4\frac{2}{3}) = 12$ $-4\frac{2}{3} = -\frac{1}{3}(-1) - 5$
−2 + 14 = 12 $-4\frac{2}{3} = \frac{1}{3} - 5$
$12 \overset{\checkmark}{=} 12$ $-4\frac{2}{3} \overset{\checkmark}{=} -4\frac{2}{3}$

Solution: x = −1 and $y = -\frac{14}{3}$

When answers have fractions, it becomes ever so important to have a graphing calculator or graph paper. Preciseness does make a difference. Unless specifically requested, substitution or elimination are the preferred ways to solve systems of equations.

149

Let us try a few problems using graphing as our technique.

1. 7 = 2y + x
    5x – 3y = 9

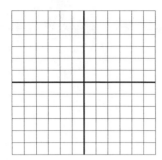

2. 2y + x = 2
    6 = x – 2y

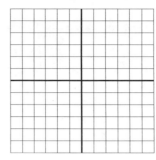

3. x + y = 9
    x – y = 3

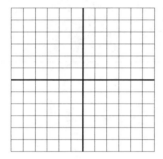

4. 2x + 2y = –8
    x – y = 4

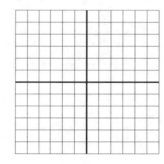

5. x – 3y = 5
    2x + y = 8

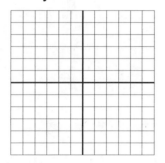

6. 2x – y = –1
    y + x = 4

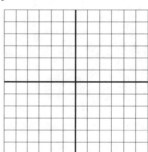

7. x + 3y = 5
    3x – 9y = 3

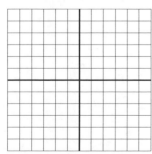

8. 2y = –3x + 6
    3x – 4y = 24

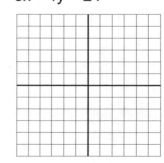

LESSON 53 – A Review of Lessons 50-52

Solve the systems of equations using any of the three methods: substitution, elimination, or graphing.

1. $3x + y = 1$
    $2x - 7 = y$

2. $x + y = 2$
    $x - y = 2$

3. $4 = 3y + 2x$
    $3x = y + 7$

4. $\dfrac{x}{3} + \dfrac{y}{2} = \dfrac{5}{6}$

    $\dfrac{x}{2} = \dfrac{5}{6} - \dfrac{y}{3}$

    hint: multiply all terms by common denominator

5. $7a - 2b = \dfrac{11}{2}$
    $4a + 5b = -3$

6. $a + 3b = -6$
    $3a = 3b - 7$

7. $4m + 2n = 0$
    $m - n = 6$

8. $a = \dfrac{4b - 3}{3}$
    $3a + 2b = 9$

9. $-4a + 4b = 5$
    $-6a - 8b = 11$

10. $2x - 5y = -4$
    $x + y = 5$

11. $3x + 4t = 18$
    $x - 3t = -7$

12. $x = 11 - 2y$
    $x - 5 = y$

LESSON 54 – Story Problems
Distance, Rate, Time

Example 1: A car4 travels 55 miles per hour (mph) for 10 hours. How far did it travel?

Solution: Distance = Rate • Time   D = R • T

$$D = \frac{55 \text{ miles}}{\text{hours}} \cdot \frac{10 \text{ hours}}{1} = 550 \text{ miles}$$

Example 2: A car travels 30 mph faster than a truck. Both vehicles leave from the same location and travel in opposite directions. After 5 hours and 500 miles between them, what is the rate of the car and the truck?

Solution:

	Rate	•	Time	=	Distance
Car	x	•	5	=	5x
Truck	x – 30	•	5	=	5(x – 30)

$$
\begin{aligned}
(\text{Distance of Car}) + (\text{Distance of Truck}) &= \text{Total Distance} \\
5x + 5(x - 30) &= 500 \\
5x + 5x - 150 &= 500 \\
10x - 150 &= 500 \\
10x &= 650 \\
x &= 65 \text{ mph}
\end{aligned}
$$

Rate of car: 65 mph
Rate of truck: 35 mph

Example 3: How long will it take a car, traveling at 55 mph, to pass a truck, which has a 3-hour head start and travels at 40 mph?

Solution:

	Rate	•	Time	=	Distance
Car	55	•	x	=	55x
Truck	40	•	x + 3	=	40(x + 3)

(Distance of Car) = (Distance of Truck)

$$55x = 40(x + 3)$$
$$55x = 40x + 120$$
$$15x = 120$$
$$x = 8 \text{ hours}$$

The car must travel 8 hours to pass the truck.

Example 4. Two trains leave Union Station at 2:30 PM traveling in opposite directions. At 7 PM, how far apart are they if Train A's speed is 25 mph faster than Train B, which travels at 40 mph?

	Rate	•	Time	=	Distance
Train A	65	•	4.5	=	65(4.5)
Train B	40	•	4.5	=	40(4.5)

$$\text{Total Distance (D)} = \text{(Distance of Train A)} + \text{(Distance of Train B)}$$
$$D = 65(4.5) + 40(4.5)$$
$$D = 282.5 + 180$$
$$D = 472.5 \text{ miles}$$

Example 5: A plane travels from Baltimore to Chicago in 2 hours. The return trip takes 1½ hours. The distance between the two cities is 1200 miles. Find the speed of the plane and the speed of the wind.

Solution: $D = R \cdot T$

Let x = speed of the plane
Let y = speed of the wind

$$(x + y) \cdot 1\tfrac{1}{2} = 1200 \quad \text{(A)} \qquad \text{Chicago to Baltimore}$$
$$(x - y) \cdot 2 = 1200 \quad \text{(B)} \qquad \text{Baltimore to Chicago}$$

Divide equation (A) by $1\tfrac{1}{2}$ and equation (B) by 2.

$$x + y = 800$$
$$x - y = 600$$

Solve by elimination:

$$2x = 1400$$
$$x = 700 \qquad \text{speed of plane}$$

Substitute x = 700 in equation (B).

$$(x - y) \cdot 2 = 1200$$
$$(700 - y) \cdot 2 = 1200$$
$$1400 - 2y = 1200$$
$$-2y = -200$$
$$y = 100 \text{ mph} \qquad \text{speed of wind}$$

Check:

Equation A	Equation B
$(x + y) \cdot 1\frac{1}{2} = 1200$	$(x - y) \cdot 2 = 1200$
$(700 + 100) \cdot 1\frac{1}{2} = 1200$	$(700 - 100) \cdot 2 = 1200$
$(800) \cdot 1\frac{1}{2} = 1200$	$(600) \cdot 2 = 1200$
$1200 \overset{\checkmark}{=} 1200$	$1200 \overset{\checkmark}{=} 1200$

Solution: speed of plane: 700 mph
speed of wind: 100 mph

Exercises.

1. Two trucks pass each other along I-95 and continue traveling in opposite directions. One truck is traveling at 40 mph, and the other at 50 mph. How many hours does it take for both the trucks to be 225 miles apart?

2. At noon, two airplanes leave an airport and travel in opposite directions at the same altitude. The slower plane is flying at 360 mph and the faster at 450 mph. At what time will they be 1350 miles apart?

3. A motorboat home traveled upstream on a river at 15 mph and returned downstream at 20 mph. How far upstream did the motorboat home travel if the round trip took $3\frac{1}{2}$ hours?

4. Two walkers start at the same time from opposite ends of a boardwalk that is 12 kilometers long. One walks twice as fast as the other. Find the rate, in km/h, of each walker if they meet in 15 minutes. (hint: 15 minutes = $\frac{1}{4}$ hour)

5. A train that averages 100 km/h can travel from New York to Albany in 2 hours less time than a motorist driving at an average speed of 75 km/h. Find the distance from New York to Albany.

6. A round trip by bus between Ocean Lake and Springtown took 3 hours. If the bus averaged 40 mph in one direction and 50 mph in the other, find the distance between Ocean Lake and Springtown.

7. Two jets leave Pittsburgh at the same time, one flying due north at 900 km/h, and the other flying due south at 1050 km/h. After how many hours will the planes be 3900 km apart?

8. Mr. Brown lives 15 miles from the city in which he works. He walks for 15 minutes at 4 mph and then boards a bus that takes him to the city in 20 minutes. Find the speed of the bus.

9. Bill and Sarah were each bicycling from Allentown to Clark Summit at 10 AM. Sarah was 3 miles ahead of Bill. Sometime later, Bill passed Sarah. At what time did Bill pass Sarah if Sarah's speed was 15 mph and Bill's was 19 mph?

10. A cyclist left Springford at 10 AM and traveled due west at 17 mph. At 11 AM, a motorist left Springford and traveled the same route as the cyclist at 47 mph. At what time did the motorist overtake the cyclist?

11. A passenger train and a freight train start from the same point at the same time and travel in opposite directions. The passenger train travels four times as fast as the freight train. In 6 hours, they are 660 miles apart. Find the speed (rate) of each train.

12. Two airplanes started at the same time from airports 1500 miles apart and flew towards each other. They passed each other in 3 hours. The faster plane averaged 40 mph more than the slower plane. Find the rate and distance each plane flew.

13. Buster has a motor boat that can travel 15 mph in still water. He wishes to make a trip on a river whose current flows at a rate of 3 mph. If Buster has 8 hours to motor his boat, how far can he travel up the river and return?

14. Two automobiles make the same trip of 300 miles. One traveled 20 mph faster than the other and required 4 less hours to make the trip. Find the rate of each automobile.

LESSON 55 – Story Problems
Coins

Example 1: Fifty quarters and dimes have a total value of $11. How many quarters and dimes are there?

Solution:

Let q = number of quarters
Let d = number of dimes

0.25 = value of a quarter
0.10 = value of a dime

$$q + d = 50 \qquad (A)$$
$$0.25q + 0.10d = 11 \quad (B)$$

Multiply equation (B) by 100 to remove the decimals.

$$q + d = 50 \qquad (A)$$
$$25q + 10d = 1100 \quad (B)$$

Solve by elimination. Multiply equation (A) by –10.

$$-10q - 10d = -500$$
$$25q + 10d = 1100$$

Solve for "q".

$$15q = 600$$
$$q = 40$$

Substitute q = 40 in equation (A).

$$q + d = 50$$
$$40 + d = 50$$
$$d = 10$$

Check: $40(0.25) + 10(0.10) = 11$
$$10 + 1 = 11$$
$$11 \overset{\checkmark}{=} 11$$

Solution: 40 quarters and 10 dimes

Example 2:   Jake has some nickels and quarters. He has 33 coins altogether with $3.75 more in quarters than nickels. How many quarters and nickels does he have?

Solution:

$$n + q = 33 \qquad \text{(A)}$$
$$0.25q = \$3.75 + 0.05n \qquad \text{(B)}$$

Multiply equation (B) by 100 to remove the decimals.

$$n + q = 33 \qquad \text{(A)}$$
$$25q = 375 + 5n \qquad \text{(B)}$$

Solve by substitution. Solve equation (A) for "n".

$$n = 33 - q$$

Substitute in equation (B).

$$25q = 375 + 5n$$
$$25q = 375 + 5(33 - q)$$
$$25q = 375 + 165 - 5q$$
$$30q = 540$$
$$q = 18$$

Substitute q = 18 in equation (A).

$$n + q = 33$$
$$n + 18 = 33$$
$$n = 15$$

Check: $n + q = 33$ $0.25q = 3.75 + 0.05n$
              $15 + 18 = 33$ $0.25(18) = 3.75 + 0.05(15)$
              $33 \overset{\checkmark}{=} 33$ $4.5 = 3.75 + 0.75$
                                    $4.5 \overset{\checkmark}{=} 4.5$

Solution: 18 quarters and 15 nickels

Exercises

1. Sixty dimes and quarters were emptied from a candy machine. Their total value was $9.75. How many of each were there?

2. Susan saved nickels and dimes. If the 50 coins had a value of $3.90, how many of each type did she save?

3. There are 25 coins in a collection of nickels and quarters. If there are 3 more nickels than quarters, how much are the 25 coins worth?

4. Burgess has $2.05 in dimes and nickels. There are 4 more dimes than nickels. How many dimes does he have?

5. Roland has $2.15 in dimes and quarters. The number of dimes is 2 more than four times the number of quarters. How many quarters and dimes does he have?

6. Timothy only collects quarters and half-dollars. Currently his collection is worth $28.75. He has five more half-dollars than quarters. How many coins does he have?

7. Luke has a collection of nickels, dimes and quarters. He has twice as many quarters as dimes. The jar contains 15 more nickels than dimes. He counts his dimes and comes up with 32. How much money is in his coin jar?

8. There are 75 more dimes than nickels in a coin jar. If the value of the dimes is $11.20, how many nickels are in the coin jar?

9. Pedro has three times as many half-dollars as quarters. The total number of coins is 32. How much money is in the coin jar?

10. The value of dimes and quarters is $4. If we interchanged the coins, their value would by $5.80. How many of each are there?

LESSON 56 – Story Problems
Digits

Example 1: The sum of the digits of a two-digit number is 9. The number is 12 times the tens digit. What is the number?

Solution:

Let t = tens digit
Let u = ones digit
number = 10t + u

$t + u = 9$ (A)
$10t + u = 12t$ (B)

Use substitution. Solve equation (A) for "u" and substitute into equation (B). Solve for "t"

$t + u = 9$
$u = 9 - t$

$10t + u = 12t$
$10t + (9 - t) = 12t$
$9t + 9 = 12t$
$9 = 3t$
$3 = t$

Substitute t = 3 into equation (A). Solve for "u".

$t + u = 9$
$3 + u = 9$
$u = 6$

Check the solution.

Equation (A)
   $t + u = 9$
   $3 + 6 = 9$
   $9 \overset{\checkmark}{=} 9$

Equation (B)
   $10t + u = 12t$
   $10(3) + 6 = 12(3)$
   $30 + 6 = 36$
   $36 \overset{\checkmark}{=} 36$

Solution: The number is 36.

Example 2:   The sum of the digits of a two-digit number is 15. The tens digit is one-half of three times the units digit. What is the number?

Solution:

Let t = tens digit
Let u = ones digit
number = 10t + u

$t + u = 15$ (A)

$t = \dfrac{3u}{2}$ (B)

Solve by substitution. Substitute equation (B) into equation (A).

$\dfrac{3u}{2} + u = 15$

Multiply all three terms by 2. Then solve for "u".

$2 \cdot \dfrac{3u}{2} + 2 \cdot u = 2 \cdot 15$
$3u + 2u = 30$
$5u = 30$
$u = 6$

Substitute u = 6 in equation (A).

$t + u = 15$
$t + 6 = 15$
$t = 9$

Check: $t + u = 15$ $t = \dfrac{3u}{2}$

                 $9 + 6 = 15$ $9 = \dfrac{3 \cdot 6}{2}$

                 $15 \overset{\checkmark}{=} 15$ $9 = \dfrac{18}{2}$

                                            $9 \overset{\checkmark}{=} 9$

The solution is 96.

Always be sure to check you solution!

Exercises

1. The sum of the digits of a two-digit number is 11. If six times the units digit is 5 less than the number, find the number.

2. The units digit of a two-digit number is two more than the tens digit. The sum of the digits is 12. Find the number.

3. The tens digit of a two-digit number is three more than the units digit. Twice the units digit is two more than the tens digit. Find the number.

4. The sum of the digits of a two-digit number is 12. Reverse the digits and the number is decreased by 54. Find the number.

5. The tens digit is zero in a three-digit number and the sum of the digits is 12. If the three-digit number value is 51 times the units digit, find the number.

6. A two-digit number equals 14 times its tens digit. The sum of the digits is 10. Find the number.

7. A two-digit number equals four times the sum of the digits. If the tens digit is two less than the units digit, find the number.

8. The sum of the digits of a three-digit number is 14, while the tens digit is 6. When the digits are reversed, the new number is 198 more than the original number. Find the original number.

9. The sum of the digits of a two-digit number is 9. When 63 is subtracted from the number, the new number is the same as the original number but the digits are reversed. Find the original number.

10. The sum of the digits of a two-digit number is 9. The number is seven times the sum of the digits. Find the number.

LESSON 57 – Story Problems
Business

Example 1:   Six bars of soap and five cans of cleanser cost $3.50 At that same store, three cans of cleanser and two bars of soap cost $1.70. How much does one bar of soap and one can of cleanser cost?

Solution:

Let s = soap
Let c = cleanser

$6s + 5c = 3.50$ (A)
$2s + 3c = 1.70$ (B)

Solve by elimination. Multiply equation (B) by (–3).

$-3(2s + 3c = 1.70)$
$-6s - 9c = -5.10$

Eliminate "s" and solve for "c".

$$6s + 5c = 3.50$$
$$\underline{-6s - 9c = -5.10}$$
$$-4c = -1.60$$
$$c = 0.40$$

Substitute "c" into equation (B). Solve for "s".

$2s + 3c = 1.70$
$2s + 3(0.40) = 1.70$
$2s + 1.20 = 1.70$
$2s = 0.50$
$s = 0.25$

Check: $6s + 5c = 3.50$ $2s + 3c = 1.70$
$6(0.25) + 5(0.40) = 3.50$ $2(0.25) + 3(0.40) = 1.70$
$1.50 + 2.00 = 3.50$ $0.50 + 1.20 = 1.70$
$3.50 \overset{\checkmark}{=} 3.50$ $1.70 \overset{\checkmark}{=} 1.70$

Solution: one can of cleanser: $0.40
one bar of soap: $0.25

Example 2:   A coat sells for $70. What is the store's cost if the selling price represents a 40% profit?

        Solution: Let x = cost of the coat

$$x + 40\%(x) = 70$$
$$x + 0.40x = 70$$
$$1.40x = 70$$
$$x = 50$$

        Check: $$x + 40\%(x) = 70$$
$$50 + 0.40(50) = 70$$
$$50 + 20 = 70$$
$$70 \overset{\checkmark}{=} 70$$

The store's cost of the coat is $50.

Example 3:   A farmer bought a number of sheep for $440. After 5 died, he sold the remaining sheep at a $2 profit each. However, he made a $60 profit for his dealings. How many sheep did the farmer originally buy?

        Solution: Let n = number of sheep
                        Let r = cost of each sheep

$$n \cdot r = 440 \qquad \text{(A)}$$
$$(n - 5)(r + 2) = 500 \quad \text{(B)}$$

Using FOIL, expand equation (B).

$$nr - 5r + 2n - 10 = 500 \qquad \text{(C)}$$

Substitute "nr" in equation (C). Simplify.

$$440 - 5r + 2n - 10 = 500$$
$$-5r + 2n + 430 = 500$$
$$2n - 5r - 70 = 0 \qquad \text{(D)}$$

Solve equation (A) for "n".

$$n \cdot r = 440$$
$$n = \frac{440}{r}$$

163

Substitute "n" in equation (D).

$$2n - 5r - 70 = 0$$

$$2\left(\frac{440}{r}\right) - 5r - 70 = 0$$

Multiply all terms by "r" to remove the fraction.

$$r \cdot 2\left(\frac{440}{r}\right) - r \cdot (5r) - r \cdot (70) = r \cdot 0$$

$$2(440) - 5r^2 - 70r = 0$$
$$880 - 5r^2 - 70r = 0 \qquad \text{Multiply by } -1.$$
$$5r^2 + 70r - 880 = 0 \qquad \text{Divide by 5.}$$
$$r^2 + 14r - 176 = 0$$

Solve the quadratic by factoring.

$$r^2 + 14r - 176 = 0$$
$$(r + 22)(r - 8) = 0$$

$$r + 22 = 0 \qquad\qquad r - 8 = 0$$
$$r = -22 \qquad\qquad\quad r = 8$$

Since "r" represents the cost of each sheep, −22 is not a possible solution. Therefore the cost of each sheep is $8.

Substitute r = 8 in equation (A).

$$n \cdot r = 440$$
$$n \cdot 8 = 440$$
$$n = 55$$

The farmer originally bought 55 sheep.

Exercises
1.   Six pounds of peas and five pounds of beans cost $1.48. At the same place and time, eight pounds of peas and seven pounds of beans cost $2.00. Find the cost of one pound of peas and one pound of beans.

2.   A furniture store paid $55 for a table. At what price should the store price the table to sell it at a discount of 20% and still make a 25% profit on the marked price?

3. A farmer bought some heifers for $720. If they had cost him $15 apiece less, he could have bought four more heifers for the same amount. How many did he buy?

4. A merchant bought a number of cans of chick peas for $14.40. Later the price increased by 2¢ per can. He found that he received 24 fewer cans for the same amount of money. How many cans of chick peas were in the original purchase?

5. Eight carnations and nine roses cost $3.35. At the same time and location, a dozen carnations and five roses cost $3.75. Find the cost of a single rose and a single carnation.

6. A group of men bought a boat for $300. If there had been twice as many men, each would have paid $30 less. How many men bought the boat?

7. Nine fountain pens and eight pencils cost $7.10. At that same time and location, six pens and four pencils cost $4. Find the cost of one pen and one pencil.

8. Sending a telegram from one location to another is 41¢ for 14 words and the cost of sending another of 20 words is 65¢. The first ten words are a fixed amount and each additional words costs extra. Find the fixed cost of ten words and the cost of each additional word.

9. The senior class of Darien Lake High School purchased a gift for their school. The cost was $400. During the year, ten seniors moved, causing an increase of $2 in the amount each senior had to pay. How many seniors were originally in the class?

10. Ten cans of soup and four cans of spaghetti cost $14.70. At that same store and day, eight cans of soup and five cans of spaghetti cost $13.92. Find the cost of one can of soup and one can of spaghetti.

11. A lady bought a certain stock for $528. If she had bought the stock when each share was $2 less, she could have bought two more shares. How many shares did she buy and at what price?

12. It costs $180 to manufacture a radio. At what price should the manufacturer mark it so that a company can sell it at a 10% discount on the marked price and still make a profit of 20% on the selling price?

13. Two cans of beans and four cans of soup cost $5.26. At that same store, four cans of beans and five cans of soup cost $7.55. What is the cost of a can of beans and a can of soup?

14. As a clothing store proprietor, you bought a man's suit for $100. You sell it at a 25% discount and still make a 50% profit. What should you sell the suit for now?

LESSON 58 – Story Problems
Age

Example 1:   Jack is 20 years older than Felix. Five years ago, Jack was five times as old as Felix was then. Find the present age of Jack and Felix.

Solution:

Let x = Felix's present age
       $x + 20$ = Jack's present age
       $x - 5$ = Felix's age five years ago
       $x + 15$ = Jack's age five years ago

Write an equation and solve.

Jack's age five years ago = 5 • (Felix's age five years ago)

$x + 15 = 5(x - 5)$
$x + 15 = 5x - 25$
$40 = 4x$
$10 = x$

Solution: Felix's present age is 10 and Jack's present age is 30.

Example 2:   Susan is now eight years old and Michelle is two years old. In how many years will Susan be twice as old as Michelle?

Solution:

	Present Age	Age in "x" Years
Susan	8	$8 + x$
Michelle	2	$2 + x$

$8 + x = 2(2 + x)$
$8 + x = 4 + 2x$
$4 = x$

Check:
$8 + x = 2(2 + x)$
$8 + 4 = 2(2 + 4)$
$12 = 2(6)$
$12 \overset{\checkmark}{=} 12$

Solution: In 4 years, Susan will be twice as old as Michelle.

Exercises

1. Wally is three times as old as Beaver. Ten years from now, Wally will be twice as old as Beaver will be then. How old is each now?

2. The sum of a man's age and his daughter's age is 50 years. Eight years from now, the man will be two times as old as his daughter will be then. How old is each now?

3. Morgan is sixteen years older than Abbie. Four years ago, Morgan was twice as old as Abbie was then. Find the present age of each.

4. The sum of Mickey's and Matt's age is 28 years. Mickey's age one year from now will be seven times Matt's age one year ago. Find the present age of both Mickey and Matt.

5. Barney is 32 years old and Duncan is 14 years old. How many years ago was Barney four times as old as Duncan?

6. Sir Hillary is 55 years old and his wife, Brigit, is 45 years old. How many years ago was the ratio of Hillary to Brigit ages is 4:3?

7. Six years ago, Sally was $\frac{1}{10}$ as old as Christine. Nine years from now, Sally will be $\frac{2}{5}$ as old as Christine. Find their present age.

8. Mrs. Hall is 15 years older than Mrs. Ryan. Five years from now, Mrs. Hall will be $1\frac{1}{2}$ times as old as Mrs. Ryan. How old is each woman now?

9. Donovan is 30 years old and Kevin is 15 years old. In how many years will Donovan be $1\frac{1}{2}$ times as old as Kevin is then?

10. Five years from now, a father will be three times as old as his son. Four years ago the father was 30 years older than his son. What is the present age of each?

LESSON 59 – Story Problems
Mixture

Example 1:   A solution of oil and gasoline is 8% oil. How much gasoline must be added to 3 gallons of the solution to obtain a new solution that is 5% oil?

Solution:

	Percent of oil (%)	×	Amount of mixture (gallons)	=	Amount of oil
original solution	8	×	3	=	0.08(3)
gasoline added	0	×	x	=	0
new solution	5	×	3 + x	=	0.05(3 + x)

$0.08(3) + 0 = 0.05(3 + x)$

$0.08(3) = 0.05(3 + x)$ multiply both sides by 100

$8(3) = 5(3 + x)$

$24 = 15 + 5x$

$9 = 5x$

$\dfrac{9}{5} = x$

Check it: $\dfrac{0.24}{3 + 1.8} = \dfrac{0.24}{4.8} = 0.05 \times 100 = 5\%$

Solution: $\dfrac{9}{5} = 1.8$ gallons of gasoline

Example 2:   How many pounds of nuts worth 45¢ per pound must be mixed with 20 pound of nuts worth 60¢ per pound to make a mixture which can be sold at 50¢ per pound?

Solution:

	Number of pounds	×	Cost per pound (¢)	=	Value of mixture
cheap nuts	x	×	45	=	45x
better nuts	20	×	60	=	20(60)
mixture	x + 20	×	50	=	50(x + 20)

$45x + 20(60) = 50(x + 20)$

$45x + 1200 = 50x + 1000$

$200 = 5x$

$40 = x$

Check: $45x + 20(60) = 50(x + 20)$
$45(40) + 20(60) = 50(40 + 20)$
$1800 + 1200 = 2000 + 1000$
$3000 \overset{\checkmark}{=} 3000$

Solution: Need 40 pounds of 45¢ nuts.

Example 3:   How much pure copper must be added to 150 pounds of an alloy which is 40% copper to produce an alloy which is 50% copper?

Solution:

	Number of pounds	×	Part pure copper (%)	=	Number of pounds of pure copper
original alloy	150	×	40	=	0.40(150)
pure copper added	n	×	100	=	1.00n
new alloy	150 + n	×	50	=	0.50(150 + n)

$0.40(150) + 1.00n = 0.50(150 + n)$
$40(150) + 100n = 50(150 + n)$ multiply both sides by 100
$6000 + 100n = 7500 + 50n$
$50n = 1500$
$n = 30$

Check: $0.40(150) + 1.00n = 0.50(150 + n)$
$0.40(150) + 1.00(30) = 0.50(150 + 30)$
$60 + 30 = 0.50(180)$
$90 \overset{\checkmark}{=} 90$

Solution: 30 pound of pure copper must be added.

Example 4:   A chemist has 160 pints of a solution which is 20% acid. How much water must she evaporate to make a solution which is 40% acid?

Solution:

	Number of pints	×	Part pure acid (%)	=	Number of pints of pure acid
original solution	160	×	20	=	0.20(160)
water evaporated	x	×	0	=	0(x)
new solution	160 − x	×	40	=	0.40(160 − x)

$0.20(160) + 0(x) = 0.40(160 − x)$
$20(160) + 0 = 40(160 − x)$ multiply both sides by 100

$$3200 = 6400 - 40x$$
$$40x = 3200$$
$$x = 80$$

Check:
$$0.20(160) + 0(x) = 0.40(160 - x)$$
$$0.20(160) + 0(80) = 0.40(160 - 80)$$
$$32 + 0 = 0.40(80)$$
$$32 \stackrel{\checkmark}{=} 32$$

Solution: 80 pints of water needs to be evaporated.

Exercises

1. A 20-gram solution of alcohol and water is 10% alcohol. How much water must be added to produce a 5% solution?

2. A farmer has two types of oats: buckwheat is 20% oats, while rolled oats is 28% oats. How many pounds of each kind should be mixed to produce 50 pounds of feed that is 25% oats?

3. How many ounces of salt must be added to 84 ounces of a 20% saltwater solution to produce a 50% saltwater solution?

4. A 6% copper alloy is to be melted with an 18% copper alloy to form a 10% copper alloy weighing 75 pounds. How much of each type must be used?

5. Candy costing 60¢ per pound is to be mixed with candy costing 85¢ per pound. Find the number of pounds of each candy to produce a mixture of 40 pounds to be sold for 70¢ per pound.

6. How many gallons of cream which contains 76% butterfat must be mixed with 18 gallons of milk which contains 12% butterfat to produce a mixture containing 60% butterfat?

7. An auto mechanic has 35% and 50% antifreeze solutions. How many quarts of each type should be mixed to produce 16 quarts of a 45% solution?

8. How many pounds of nuts worth 55¢ per pound must be mixed with 36 pounds of nuts worth 80¢ per pound to produce a mixture worth 75¢ per pound?

9. How many quarts of a solution which is 75% acid must be mixed with 16 quarts of a solution which is 30% acid to produce a solution which is 55% acid?

10. A mixture of peanuts and cashews weighs 12 ounces. Twenty-five percent of this weight is in cashews. If 2 ounces of peanuts are eaten, what percent of the remaining mixture is cashews?

11. A chemist has 15 grams of an alcohol solution of which 12% is alcohol. How much pure alcohol must be added to obtain a 20% alcohol solution?

12. A druggist has a 35% acid solution and a 45% acid solution. How many grams of each must be mixed to form 80 grams of solution that is 40% acid?

13. A milk plant has 26% butterfat-cream and 36% butterfat-cream. How many pounds of each must be mixed to produce 280 pounds of 30% butterfat-cream?

14. How many quarts of pure antifreeze must be added to 6 quarts of a 40% antifreeze solution to obtain a 50% antifreeze solution?

15. The candy counter in a department store has chocolates that sell for $1.65 a pound and caramels that sell for $1.25 a pound. A mixture of these candies is to sell for $1.49 a pound. How many pounds of each kind of candy should be used to make 10 pounds of the $1.49 mixture?

LESSON 60 – Story Problems
Work

Example 1: Joan can paint her bedroom in six hours. With help from her brother, Joe, they paint that same bedroom in two hours. How long does it take Joe to paint the bedroom working alone?

Solution: Let "x" be the time required for Joe to paint the bedroom.

Joan's time + Joe's time = Time working together

$$\frac{1}{6} + \frac{1}{x} = \frac{1}{2}$$

Multiply all three terms by the LCD, which is 6x.

$$\frac{\cancel{6} \times 1}{\cancel{6}} + \frac{6 \ast 1}{\cancel{x}} = \frac{\overset{3}{\cancel{6}} \times 1}{\cancel{2}}$$

$x + 6 = 3x$
$6 = 2x$
$3 = x$

Solution: It takes Joe 3 hours to paint the bedroom.

Example 2: Bob mows the yard at his house in seven hours. When his son, Jake, does the mowing, he only takes five hours. If they work together to mow the yard, how long will it take them?

Solution: Let "x" represents working together.

$$\frac{1}{7} + \frac{1}{5} = \frac{1}{x}$$

Multiply all three terms by the LCD, 35x (7 • 5 • x).

$$\frac{\overset{5}{\cancel{35}} \times 1}{\cancel{7}} + \frac{\overset{7}{\cancel{35}} \ast 1}{\cancel{5}} = \frac{35 \ast 1}{\cancel{x}}$$

$5x + 7x = 35$
$12x = 35$
$x = \dfrac{35}{12} = 2\frac{11}{12} = 2$ hrs 55 min

Solution: It takes 2 hrs 55 min to mow the yard together.

Example 3: If the intake pipe of a tank can fill it in three hours while it takes eight hours to drain the tank, how long will it take to fill the tank if both intake and drain pipes are open?

Solution:

$$\frac{1}{3} - \frac{1}{8} = \frac{1}{x}$$

Multiply by the common denominator: 24x

$$\frac{^8\cancel{24} \times 1}{\cancel{3}} - \frac{^3\cancel{24} \times 1}{\cancel{8}} = \frac{24 \times 1}{\cancel{x}}$$

$8x - 3x = 24$
$5x = 24$
$x = \dfrac{24}{5} = 4\frac{4}{5} = 4.8 = 4$ hours 48 minutes

Check:

$$\frac{1}{3} - \frac{1}{8} = \frac{1}{\frac{24}{5}}$$

$$\frac{8}{24} - \frac{3}{24} = \frac{5}{24}$$

$$\frac{5}{24} \overset{\checkmark}{=} \frac{5}{24}$$

$$\frac{1 \times \frac{5}{24}}{\frac{24}{5} \times \frac{5}{24}} = \frac{\frac{1}{1} \times \frac{5}{24}}{\frac{24}{5} \times \frac{5}{24}} = \frac{\frac{5}{24}}{1} = \frac{5}{24}$$

Solution: It takes 4 hours 48 minutes to fill the tank.

Exercises

1. Marcus can mow his neighbor's grassy yard in four hours. His friend, Jon, can mow that same yard in three hours. How long will it take both men to mow the yard working together?

2. A cleaning service can clean a two-story house in two hours. If the owner cleans the house, it takes five hours. How long will it take both the owner and cleaning service to clean the two-story house?

3. Mark and Mike can paint the outside of a house in five hours. When Mark paints it alone, it takes him nine hours. How long will Mike take to paint the house if he works alone?

4.   Tom can sand a floor in four hours. Kim takes six hours to sand that same floor. If they both work together, how long will it take them?

5.   A truck driver can load his truck in five hours. With help from a friend, it takes only $2\frac{1}{2}$ hours. How long does the friend take to load the truck alone?

6.   One pipe can fill a tank in one hour. Another pipe can fill that same tank in 48 minutes. How long will it take both pipes to fill the tank?

7.   A drain pipe can remove all the water from a pool in six hours. It takes the intake pipe four hours to fill the tank. If both pipes are open, how long does it take to fill the pool?

8.   A farmer can plow a field with a two-prong plow in five hours. If his brother uses a three-prong plow, he can do the plowing in three hours. If the farmer gets his brother to help, how long will it take to plow the field?

9.   Damian can paint a room in three hours. His cousin can paint the same room in two hours. After Damian paints for one hour, his cousin joins him. How long will it take to complete the painting?

10. Two inlet pipes can fill a pool in six hours. The larger pipe can fill the pool by itself in one-third the time it takes the smaller pipe. How long does it take each pipe alone to fill the pool?

11. It takes a man 8 hours to do a certain job and a boy 12 hours to do the same job. If 2 men and 3 boys work on the job, how many hours does it take to finish it?

12. Henry and Clyde can build a brick wall in 4 days if the work together. If Henry would need 6 days more than Clyde to build the wall alone, how many days would each individual working alone need for the job?

LESSON 61 – Story Problems
Investment

Example 1: $500 is deposited at a local savings bank that pays $4\frac{1}{4}$% interest per year. How much interest is earned in six months? Round answer to the nearest cent.

Solution: I = PRT Interest = Principal × Rate × Time

Put all percents over 100: $4\frac{1}{4}\% = 4.25\% = \dfrac{4.25}{100}$

Since the rate is per year, convert 6 months to part of a year: 6 months = $\frac{1}{2}$ year.

$$I = \$500 \times 4.25\% \times 6 \text{ months}$$

$$I = 500 \times \frac{4.25}{100} \times \frac{1}{2}$$

$$I = \frac{\overset{5}{\cancel{500}}}{1} \times \frac{4.25}{\underset{1}{\cancel{100}}} \times \frac{1}{2} = \frac{21.25}{2} = 10.625 = 10.63$$

Solution: $10.63 interest in 6 months

Example 2:   If $20,000 is invested at two different interest rates, 5% and 6%, and $1120 interest is received from both rates, how much is invested at each rate?

Solution: (Interest at 5%) + (Interest at 6%) = (Total Interest)

Let "x" = amount invested at 5%.
Let "20,000 – x" = amount invested at 6%.

5%(x) + 6%(20,000 – x) = 1120
0.05x + 0.06(20,000 – x) = 1120   multiply both sides by 100
5x + 6(20,000 – x) = 112,000
5x + 120,000 – 6x = 112,000
–x = –8000
x = 8000
20,000 – x = 12,000

Check: 5%(x) + 6%(20,000 – x) = 1120
400 + 720 = 1120
1120 $\overset{\checkmark}{=}$ 1120

Solution: $8000 at 5% and $12,000 at 6%

Example 3: How long does it take a $5000 investment to earn $860 interest if the rate is 8% and not compounded?

Solution: PRT = I

$$\$5000 \times 8\% \times T = \$860$$

$$\frac{\overset{50}{\cancel{5000}}}{1} \times \frac{8}{\underset{1}{\cancel{100}}} \times T = 860$$

$$50 \times 8 \times T = 860$$

$$400T = 860$$

$$T = \frac{860}{400} = \frac{43}{20} = 2\frac{3}{20}$$

Check: $$\$5000 \times 8\% \times T = \$860$$

$$5000 \times \frac{8}{100} \times 2\frac{3}{20} = 860$$

$$5000 \times \frac{2}{25} \times \frac{43}{20} = 860$$

$$200 \times 2 \times \frac{43}{20} = 860$$

$$400 \times \frac{43}{20} = 860$$

$$20 \times 43 = 860$$

$$860 \overset{\checkmark}{=} 860$$

Solution: $2\frac{3}{20}$ years

Exercises

1. An investment of $2000 is placed in a savings account for two years. How much interest is earned, not compounded, at a $6\frac{3}{4}\%$ rate?

2. How long will it take $1000, invested at 5% not compounded, to receive $200 in interest?

3. $10,000 is invested at a $5\frac{1}{2}\%$ rate, while $15,000 is invested at a $7\frac{1}{4}\%$ rate. How much interest is earned from each investment for one year? What is the total interest income?

4. Which is the better investment: a CD (certificate of deposit) paying $4\frac{3}{4}$% monthly or 5% quarterly for one year? (no compounding)

5. How much interest is earned on $500 for 6 months at 5.60% yield?

6. A bank invested $40,000 in two small businesses. For the year, the bank earned 17% interest on one business while losing 12% on the other. The bank's net profit on these two investments was $1000. How much was invested in each business?

7. A $50,000 savings bond earns 8% interest per year. Find how much interest is earned after three months, not compounded.

8. A $1000 investment earns $4\frac{1}{2}$% interest yearly. What is the interest earned after one year? What is the interest earned if the interest after the first year is added to the principal? What is the principal after two years.

9. Three-fifths of an investment sum is invested at 8% and the remainder is invested at 10%. The total interest on both investments is $1249.60. How much is invested at each rate?

10. A sum of money is invested at an annual rate of 9% earns $1350 interest for the year. Twice this sum invested at a yearly rate of 8% earns $2400 interest for the year. How much is invested at each rate?

11. Part of an investment of $6000 is invested at 6% interest per year. The remainder is invested at 4% interest per year. Each investment earns $144 yearly interest. How much is invested at each rate?

12. One investment at 7% interest earns $980 interest per year. Half the dollar amount of the 7% investment is invested at 9% and earns $630 per year. Find the amount invested at each percent and the total investment.

LESSON 62 – Story Problems
Numbers

Example 1: The larger of two numbers is three more than twice the smaller. Their sum is 39. Find both numbers.

Solution: Let "n" = smaller number (the one you know the least about)
Let "2n + 3" = larger number

smaller number + larger number = 39
$n + 2n + 3 = 39$
$3n + 3 = 39$
$3n = 36$
$n = 12$
$2n + 3 = 2(12) + 3 = 24 + 3 = 27$

Check: $n + 2n + 3 = 39$
$12 + 2(12) + 3 = 39$
$12 + 24 + 3 = 39$
$39 \overset{\checkmark}{=} 39$

Solution: 12 and 27

Example 2: Find three consecutive even integers such that the sum of the first and third exceeds one-half of the second by 15.

Solution: Let "n" = first integer.
Let "n + 2" = second integer.
Let "n + 4" = third integer.

first integer + third integer = $\frac{1}{2}$(second integer) + 15
$n + n + 4 = \frac{1}{2}(n + 2) + 15$
$2n + 4 = \frac{1}{2}n + 1 + 15$
$2n + 4 = \frac{1}{2}n + 16$
$\frac{3}{2}n = 12$ multiply both sides by $\frac{2}{3}$ (reciprocal of $\frac{3}{2}$)
$n = 8$
$n + 2 = 10$
$n + 4 = 12$

Check: $n + n + 4 = \frac{1}{2}(n + 2) + 15$

$8 + 12 = \frac{1}{2}(10) + 15$

$20 = 5 + 15$

$20 \overset{\checkmark}{=} 20$

Solution: 8, 10, and 12

Example 3: What is the number that when added to both the numerator and denominator of $\frac{16}{34}$, the new fraction equals $\frac{4}{7}$?

Solution: Let "n" = the number.

$$\frac{16 + n}{34 + n} = \frac{4}{7} \qquad \text{cross multiply}$$

$7(16 + n) = 4(34 + n)$

$112 + 7n = 136 + 4n$

$3n = 24$

$n = 8$

Check: $\dfrac{16 + n}{34 + n} = \dfrac{4}{7}$

$\dfrac{16 + 8}{34 + 8} = \dfrac{4}{7}$

$\dfrac{24}{42} = \dfrac{4}{7}$

$\dfrac{4}{7} \overset{\checkmark}{=} \dfrac{4}{7}$

Solution: 8

Exercises
1. The larger of two numbers is two less than twice the smaller number. Their sum equals 25. Find both numbers.

2. Find three consecutive even numbers such that the sum of the two smaller numbers exceeds the third by 4.

3. Find two numbers such that their sum equals 33 and their difference equals 9.

4. What is the number when added to both the numerator and denominator of $\frac{12}{22}$ equals $\frac{11}{16}$?

5. The denominator of a fraction is five more than the numerator. If the numerator is multiplied by 3 and the denominator by 2, the resulting fraction equals $\frac{12}{13}$, find the fraction.

6. Find three consecutive even numbers such that the sum of the first two exceeds the third number by 12.

7. Find two numbers such that their sum equals 36 and their difference equals the smaller number.

8. Find two numbers such that their product equals 48 and their sum equals the smaller number squared.

9. Find two single-digit numbers such that their product equals 36 and their difference is 5.

10. Separate 58 into two parts such that the larger divided by the smaller gives a quotient of 4 and a remainder of 3.

11. One number exceeds four times a second number by 4. One-third of the larger exceeds one-half of the smaller by 8. Find both numbers.

12. What number must be added to both the numerator and denominator of the fraction $\frac{5}{23}$ to give a fraction equal to $\frac{1}{3}$?

LESSON 63 – Story Problems
Inequalities

Inequality problems usually have these key words: "at least", "no more than", "at most" and "no less than". We need to be able to translate these phrases into math symbols.

Example 1: Three added to two times a positive integer is at most 15. Find all the integers and the largest possible answer that will satisfy the criteria.

       Solution: Let "n" = integer.

                    Symbolically, "at most" is "$\leq$".

                    The inequality is: $2n + 3 \leq 15$

$$2n + 3 \leq 15 \quad \text{subtract 3}$$
$$2n \leq 12 \quad \text{divide by 2}$$
$$n \leq 6$$

                    All possible integers are: $1 \leq x \leq 6$

       Solution: $n = \{1, 2, 3, 4, 5, 6\}$ and the largest positive integer is 6.

Example 2: The sum of two consecutive integers is no less than 61. Find the smallest pair of numbers.

       Solution: Let "n" = first integer.
                        Let "n + 1" = second integer.

                    Symbolically, "no less than" is "$\geq$".

                    The inequality is: first integer + second integer ≥ 61

$$n + n + 1 \geq 61$$
$$2n + 1 \geq 61 \quad \text{subtract 1}$$
$$2n \geq 60 \quad \text{divide by 2}$$
$$n \geq 30$$

                    All pairs of consecutive integers beginning with 30.

       Solution: The smallest pair of integers is 30 and 31.

Example 3:   Three times the sum of 8 and a number is at least 93. Find the smallest possible number.

Solution: Let "n" = number.

Symbolically, "at least" is "≥".

The inequality is: $3(8 + \text{number}) \geq 93$

$$3(8 + n) \geq 93$$
$$24 + 3n \geq 93$$
$$3n \geq 69$$
$$n \geq 23$$

Solution: The smallest possible number is 23.

Example 4:   The sum of two consecutive odd numbers is less than 46. Find the pair with the largest sum.

Solution: Let "n" = first number.

Let "n + 2" = second number

Symbolically, "less than" or "no more than" is "≤"

The inequality is: $n + n + 2 \leq 46$

$$n + n + 2 \leq 46$$
$$2n + 2 \leq 46$$
$$2n \leq 44$$
$$n \leq 22$$

Note that the largest _**odd**_ number is 21. The next consecutive odd number is 23. Together, their sum (44) is less than 46.

Solution: The pair with the largest sum and fits the criteria of two consecutive odd numbers is 21 and 23.

READ EACH QUESTION CAREFULLY!!!

Exercises

1. The sum of two consecutive integers is no more than 70. Find the largest integers for which this is possible.

2. The sum of 4 and three times a number is at most 31. Find the largest possible number.

3. If 3 is added to four times an integer, their sum is less than 36. Find the largest integer for which this can occur.

4. Sylvester is 3 years older than Macy. The sum of their ages is less than 22. What is the oldest that Sylvester and Macy could be?

5. Tina received 90, 87, 94, and 84 on her first four math exams. What must she get on the last exam to have a 90 average in math?

6. Tom and Amy will share prize money of at least $400. If Amy's prize is $60 more than Tom's, what is the minimum amount each will receive?

7. Find the four greatest consecutive odd integers whose average is greater than their sum.

8. Penny has saved 40 coins consisting only of nickels and quarters. The total value of the coins is less than $5.30. What is the least number of nickels she must have?

9. The average of six consecutive integers is less than 15. What are the largest values possible for the integers?

10. Two jet aircraft take off from the same base at the same time. One flies due north at 580 mph and the other due south at 640 mph. After how many hours will they be no more than 4880 miles apart?

LESSON 64 – Story Problems
Percents

Example 1:   What is 80% of 60?

Solution: $\dfrac{Percent}{100} = \dfrac{Part}{Whole}$

Let "N" = part.

$\dfrac{80}{100} = \dfrac{N}{60}$ cross multiply

100N = 80(60)

100N = 4800

N = 48

Solution: 48

Example 2:   16 is part percent of 40?

Solution: $\dfrac{Percent}{100} = \dfrac{Part}{Whole}$

Let "N" = percent.

$\dfrac{N}{100} = \dfrac{16}{40}$ cross multiply

40N = 16(100)

40N = 1600

N = 40

Solution: 40%

Example 3:   12 is 30% of what?

Solution: $\dfrac{Percent}{100} = \dfrac{Part}{Whole}$

Let "N" = whole.

$\dfrac{30}{100} = \dfrac{12}{N}$ cross multiply

$$30N = 12(100)$$
$$30N = 1200$$
$$N = 40$$

Solution: 40

Example 4:   Tom bought an I-Pod for $300, which was a discount of 25%. What was the original price?

Solution: Rephrase the question! If the discount was 25%, how much was actually paid?

The new question would be: $300 = 75% of what number?

$$\frac{Percent}{100} = \frac{Part}{Whole}$$

Let "N" = original price.

$$\frac{75}{100} = \frac{300}{N}$$ cross multiply
$$75N = 300(100)$$
$$75N = 30000$$
$$N = 400$$

Solution: The original price is $400.

Exercises.

1.   42 is what percent of 210?

2.   Day old bread sells for 60% of its original price of $3.60. What is the cost of the bread today?

3.   How much does one save on a DVD player that originally sells for $220 and has a 25% off tag?

4. How much does a TV cost that is on sale for $169 and you have a 15% off coupon?

5. Thirty percent off a $200 coat is what?

6. How much does the coat from problem #5 cost?

7. A state tax of 7% is added to the cost of a radio selling for $153. With the tax, how much does the radio cost?

8. A family's monthly rent is 16% of their monthly income. If the rent is $400, how much is the monthly income?

9. 2.50 is 1.25% of what number?

10. The commercials for an hour TV program use 20% of the time. A sponsor gets $33\frac{1}{3}$ % of this time. How many minutes does this represent?

11. 2% of a number is 12. What is the number?

12. What is the new selling price of a jacket originally priced at $80 is now 40% off?

13. If 70% of all kangaroos are female, in a herd of 600, how many are males?

14. You purchased two pairs of shoes, one for $75 and the other for $40. Half-off sale of the lesser price is the sale today. You hand the clerk a $100 bill. How much change should you get?

15. 60% of what number is 48?

16. The price of an air conditioner was reduced from $315 to $252. Find the percent of discount.

17. 72 is what percent of 600?

18. The price of jeans is $38. At a 25% off sale, the new price is what?

19. The price of a $9750 used car is reduced 26%. What is the new sale price?

20. Star bought a calculator at a 30% discount for $42. What was the original price?

21. At a 35% discount, how much does a pair of golf shoes cost if the pair sold for 120 originally?

22. A leather jacket was selling for 75% off the original ticket price of $400. Today only, there is an additional 30% off the already discounted price. Today, what is the new price of the leather jacket?

23. The coach bus holds 46 individuals. If there are two more women than men, how many women are on that bus?

LESSON 65 – Story Problems
Geometry

Example 1: The longer side of a rectangle exceeds the shorter by six feet. If we decrease the longer side by three feet and increase the shorter side by two feet, the areas of both rectangles will be equal. Find the dimensions of the original rectangle.

Solution: $A = LW$

Let "x" = shorter side of original rectangle.

	longer side	•	shorter side	=	Area
original rectangle	$x + 6$	•	x	=	$x(x + 6)$
new rectangle	$x + 3$	•	$x + 2$	=	$(x + 3)(x + 2)$

Area of original rectangle = Area of new rectangle

$$x(x + 6) = (x + 3)(x + 2) \quad \text{expand each side}$$
$$x^2 + 6x = x^2 + 5x + 6 \quad \text{x^2 terms cancel out}$$
$$6x = 5x + 6$$
$$x = 6$$
$$x + 6 = 12$$

Solution: The dimensions of the original rectangle are 12 ft by 6 ft.

Example 2: The area of a rectangle is 28 square feet while its perimeter is 22 feet. Find the dimensions of the rectangle.

Solution: $A = LW$ (i)
             $P = 2(L + W)$ (ii)

Let "L" = length of rectangle.
Let "W" = width of rectangle.

Solve equation (ii) for "L"

$$2(L + W) = P$$
$$2(L + W) = 22 \quad \text{divide by 2}$$
$$L + W = 11 \quad \text{solve for "L"}$$
$$L = 11 - W$$

Substitute into equation (i).

$$A = LW$$
$$28 = (11 - W)(W)$$
$$28 = 11W - W^2$$
$$0 = -28 + 11W - W^2 \qquad \text{multiply through by } -1$$
$$0 = 28 - 11W + W^2$$
$$0 = (7 - W)(4 - W)$$

$$0 = 7 - W \qquad\qquad 0 = 4 - W$$
$$W = 7 \qquad\qquad\qquad W = 4$$

Substitute into equation (ii).

$P = 2(L + W)$	$P = 2(L + W)$
$22 = 2(L + 7)$	$22 = 2(L + 4)$
$22 = 2L + 14$	$22 = 2L + 8$
$8 = 2L$	$14 = 2L$
$4 = L$	$7 = L$

Solution: There are two solutions!

Length = 4 ft, Width = 7 ft
and
Length = 7 ft, Width = 4 ft

Example 3:   The width of a concrete walk around a 28 ft by 14 ft swimming pool is two feet. What is the area of this concrete walk?

Solution: We need to find the dimensions of the larger rectangle and then subtract the area of the swimming pool to arrive at the area of the concrete walk.

Area of the larger rectangle – Area of swimming pool = Area of walk

$$(28 + 4)(14 + 4) - (28)(14) =$$
$$(32)(18) - (28)(14) =$$
$$576 - 392 =$$
$$184$$

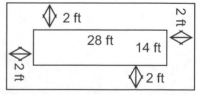

Solution: The area of the concrete walk is 184 square feet.

189

Exercises

1. The length of a rectangle exceeds its width by 12 feet. If the length is decreased by three feet and the width is increased by two feet, the areas of both rectangles would be the same. Find the dimensions of the original rectangle.

2. The length of a rectangle exceeds twice its width by 6 inches. The perimeter of the rectangle is 72 inches. Find the dimensions of the rectangle.

3. Find the dimensions of a rectangle whose perimeter is $\frac{1}{6}$ of its area.

4. The formula for the area of a triangle is $A = \frac{1}{2}bh$ where "b" is the base and "h" is the height. The area of a triangle is 36 square feet. What are the dimensions of the triangle if the base is half the height?

5. The base of an isosceles triangle is one foot larger than each of the two equal sides. If the area of the triangle is 12 sq ft and its perimeter is 16 ft, find the height and base of this isosceles triangle.

6. A picture 6 inches by 12 inches is surrounded by a frame of uniform width. If the area of the frame is twice the area of the picture, find the width of the frame.

7. The area of a square equals side squared. The perimeter equals four times the side. Find side of a square whose area and perimeter are the same.

8. If one side of a square is increased by two feet and an adjacent side is decreased by two feet, a rectangle is formed whose area is 60 square feet. Find a side of the square.

9. A rectangular lawn is 60 ft by 80 ft. How wide a uniform strip of grass must be cut around the edge when mowing in order that half of the grass be cut?

10. The length of a rectangular swimming pool is twice its width. The pool is surrounded by a cement walk 4 feet wide. The area of the cement walk is 784 square feet. Find the dimensions of the pool.

11. A picture 9 inches by 12 inches is surrounded by a frame of uniform width. If the area of the frame exceeds the area of the picture by 124 square inches, find the width of the frame.

12. The area of a rectangle is 48 square inches. If the length is increased by 4 inches and the width is increased by 2 inches, the new rectangle has twice the area of the original rectangle. Find the dimensions of the original rectangle.

13. One square has a perimeter twice that of another square. If the perimeter of the smaller square is increased by 16 inches, it would have the same perimeter as the larger square. How long are the sides of **_each_** square?

14. An isosceles triangle has two sides, each 4 inches longer than the third side. Its perimeter is 68 inches. Find the length of the three sides.

15. A rectangle has a perimeter of 30 inches. If the length were increased an amount equal to its width, the perimeter would be 42. Find the length and width of the rectangle.

16. Find the dimensions of a rectangle whose area is 72 in.^{2} and its perimeter is 34 inches.

17. The length of a rectangle exceeds twice its width by 1 inch. The perimeter is 32 inches. Find its dimensions.

18. The length of a rectangle is 5 inches less than three times its width. The perimeter is 86 inches. Find the length and width.

19. The perimeter of a right triangle is 40 inches. If the hypotenuse is 17 inches, find the length of each leg.

20. Find the dimensions of a rectangle if its perimeter is 34 feet and its diagonal is 13 feet.

LESSON 66 – A Review of Lessons 54-65

1. Two automobiles leave from Point A and travel in opposite directions. Automobile D travels at 55 mph while automobile E travels at 65 mph. After how many hours will they be 540 miles apart?

2. Luke collects quarters and nickels in his coin collection. He currently has three times as many nickels as quarters. The value of his collection is $56. How many nickels and quarters does he have?

3. The units digit of a two digit number is four times the tens digit. The sum of the two digits equals 10. What is the number?

4. Three cans of corn and 5 cans of beans cost $4.50. At that same store 5 cans of corn and 8 cans of beans cost $7.30 at the same moment in time. How much does a can of corn and a can of beans cost at that store?

5. Matt is 6 years old and half his brother Mike's age. In how many years will Matt be 2/3 of Mike's age?

6. A special mixture of nuts cost $x per pound. It consists of peanuts, costing $1.10 per pound and cashews, costing $4.50 per pound. If 20 pounds of the special mixture is 3/5 peanuts, what is the cost per pound of the special mixture?

7. Larry and Terry mow lawns for extra money. At 1234 Main St, Larry mows the lawn in 3 hours while it takes Terry 5 hours. If they work together and charge $36 per hour (which they split equally), how much will they each make when finished mowing the lawn at 1234 Main St?

8. Lilly invests $5000 at 6% for one year while her sister, Lucy, invests twice that amount at 2.9%. At the end of one year, who gets the most interest and how much more?

9. If 4 is subtracted from nine times a certain number, the result is equal to two times the square of the number. Find the number.

10. Trina received 92, 85, 88, and 93 on her first four math exams. What **_must_** she receive on her last math exam to receive at least a 90-point average and receive a B$^+$?

11. In the winter, you get a 30% reduction on a lawn mower that costs $299.50. How much do you **_save_**?

12. The area of a rectangle is 35 sq ft. If both the length and width of the rectangle were increased by 3 ft, a new rectangle is formed with an area of 80 sq ft. Find the dimensions of the original rectangle.

LESSON 67 - Functions

By definition, a set of ordered pairs or a graph is a function, if and only if, for every "x" there is a unique "y". Let us explain with examples.

Example 1:   Is the following sets of number pairs a function?

(a)   { (1, 1) , (2, 3) , (5, 7) , (9, 11) }

Since there is a unique "y" for every "x", this set of ordered pairs is a function. It can also be written this way:

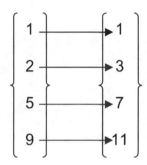

(b)   { (2, 3) , (3, 4) , (3, 5) , (6, 8) }

Because of (3, 4) and (3, 5), this set of ordered pairs is NOT a function.

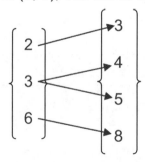

(c)   { (1, 1) , (1, 2) , (1, 3) , (1, 4) , (1, 5) }

No, this set of ordered pairs has the same "x" for five "y's". Therefore, it is NOT unique. If we were to graph these five points on a coordinate plane, it would look like the graph to the right. If we were to connect the points, a VERTICAL line would be created.

194

In all these examples, the set of ordered pairs is a relation, but examples (b) and (c) are not functions.

A Generalization: Every set of ordered pairs is a relation but not every set of ordered pairs (relation) is a function.

Example 2:   Is the equation y = 2x + 3 a function?

Solution: The equation IS a function since for every value of "x" there is a unique solution (that being "y"). Sometimes we write the equation as f(x) = 2x + 3 and it is read "f of x". If we chart this function it would look something like this:

x	2x + 3	y
0	2(0) + 3	3
1	2(1) + 3	5
2	2(2) + 3	7
3	2(3) + 3	9
−1	2(−1) + 3	1
−2	2(−2) + 3	−1
−3	2(−3) + 3	−3

No matter what number replaces "x", a unique "y" will appear. Therefore, we have a function.

Since f(x) was just introduced, we need to examine this functional notation symbolism. We may also replace f(x) with g(x), z(x), or any other letter.

$$y = x - 5$$
$$f(x) = x - 5$$
$$f(2) = 2 - 5 = -3$$
$$f(-1) = -1 - 5 = -6$$
$$f(6) = 6 - 5 = 1$$

Example 3:   Let f(x) = 2x − 5, find f(−2), f(0), f(3), and f(−6).

Solution: $f(-2) = 2(-2) - 5 = -4 - 5 = -9$
$f(0) = 2(0) - 5 = 0 - 5 = -5$
$f(3) = 2(3) - 5 = 6 - 5 = 1$
$f(-6) = 2(-6) - 5 = -12 - 5 = -17$

Example 4:   Let $g(x) = x^3 + 2x^2 - 3$, find $g(1)$, $g(-2)$, $g(2)$, and $-g(2)$.

Solution: $g(1) = (1)^3 + 2(1)^2 - 3 = 1 + 2(1) - 3 = 1 + 2 - 3 = 0$
$g(-2) = (-2)^3 + 2(-2)^2 - 3 = -8 + 2(4) - 3 = -8 + 8 - 3 = -3$
$g(2) = (2)^3 + 2(2)^2 - 3 = 8 + 2(4) - 3 = 8 + 8 - 3 = 13$
$-g(2) = -g(-2) = -[(-2)^3 + 2(-2)^2 - 3] = -[-8 + 2(4) - 3]$
$\qquad\qquad = -[-8 + 8 - 3] = -[-3] = 3$

We need to try another linear equation to show this functional property.

Example 5:   Is $y = -\frac{2}{3}x + 5$ a function?

Solution: Here is the chart and graph of the equation.

x	$-\frac{2}{3}x + 5$	y
0	$-\frac{2}{3}(0) + 5$	5
1	$-\frac{2}{3}(1) + 5$	$4\frac{1}{3}$
−1	$-\frac{2}{3}(-1) + 5$	$5\frac{2}{3}$
2	$-\frac{2}{3}(2) + 5$	$3\frac{2}{3}$
−2	$-\frac{2}{3}(-2) + 5$	$6\frac{1}{3}$
3	$-\frac{2}{3}(3) + 5$	3
−3	$-\frac{2}{3}(-3) + 5$	7
4	$-\frac{2}{3}(4) + 5$	$2\frac{1}{3}$
−4	$-\frac{2}{3}(-4) + 5$	$7\frac{2}{3}$

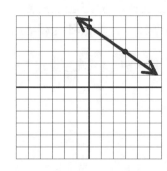

By the way, the x- and y-intercepts are as follows: (0, 5) and (7.5, 0).

The Vertical Line Test

The Vertical Line Test is another way to tell whether a graph represents a function. The depicted graph represents a function, if and only if, every vertical line intersects the graph at only one point.

The following graphs represent functions:

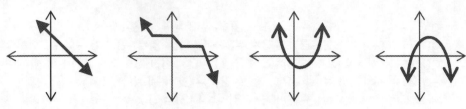

The following graphs do not represent functions:

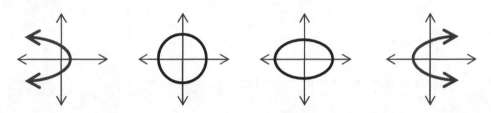

Practice.

1.   Tell whether each set of ordered pairs is a function. If not a function, explain why.

    a.   { (1, 1) , (2, 2) , (4, 4) , (6, 6) , (9, 9) }

    b.   { (2, 1) , (2, 2) , (2, 3) , (2, 4) , (2, 5) }

    c.   { (1, 0) , (2, 3) , (5, 7) , (5, 8) , (4, 9) }

    d.   { (−1, −1) , (−2, −2) , (−3, −3) , (−4, −4) , (−5, −5) }

    e.   { (−6, −3) , (−5, −2) , (−4, −1) , (−3, 0) , (−2, 1) }

    f.   { (8, 5) , (6, 6) , (7, 5) , (9, 6) , (4, 5) }

2. Tell whether each graph represents a function. Use the vertical line test in your analysis.

a. 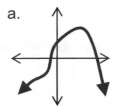   b. 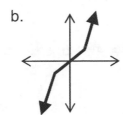   c.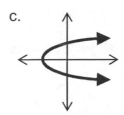

3. For the given function, find the indicated values

a. $f(x) = 3x - 1$ for   (i) $f(1)$ (ii) $f(0)$ (iii) $f(-2)$

b. $g(x) = 2x^2 - 3x + 4$ for   (i) $g(-2)$ (ii) $g(3)$ (iii) $g(-1)$

c. $h(x) = 4x^2 + 5$ for   (i) $h(-1)$ (ii) $h(2)$ (iii) $h(-3)$

d. $k(x) = -3x^2 + 6x$ for   (i) $k(1)$ (ii) $k(-2)$ (iii) $k(-1)$

e. $f(x) = x^2 + 2x - 7$ for   (i) $f(-2)$ (ii) $f(2)$ (iii) $f(-3)$

4. For each function, evaluate (i) $f(3) + f(1)$, (ii) $f(\frac{3}{4})$, (iii) $\frac{f(3)}{f(2)}$, and (iv) $f(2) - f(0)$.

a. $f(x) = 3x - 4$
b. $f(x) = x^2 + 4x$
c. $f(x) = 4x^2$
d. $f(x) = 3x^2 + 7$
e. $f(x) = 2x^2 - x + 5$

LESSON 68 – Function Within a Function and Absolute Value Functions

Several aspects of a function are discussed in this lesson. First, let us examine two distinct functions to evaluate.

Example 1:   Let $f(x) = 3x + 4$ and $h(x) = 4x - 5$. Evaluate each of the following:

a. $f(3) + h(4)$

   $f(x) = 3x + 4$ $h(x) = 4x - 5$
   $f(3) = 3(3) + 4 = 13$ $h(4) = 4(4) - 5 = 11$

   $f(3) + h(4) = 13 + 11 = \mathbf{24}$

b. $f(-2) + h(-3)$

   $f(x) = 3x + 4$ $h(x) = 4x - 5$
   $f(-2) = 3(-2) + 4 = -2$ $h(-3) = 4(-3) - 5 = -17$

   $f(-2) + h(-3) = -2 + (-17) = \mathbf{-19}$

c. $f(x) - h(x)$

   $f(x) = 3x + 4$ $h(x) = 4x - 5$

   $(3x + 4) - (4x - 5) = 3x + 4 - 4x + 5 = \mathbf{-x + 9}$

d. $f(h(2))$

   $f(x) = 3x + 4$ $h(x) = 4x - 5$

   $f(h(2)) = f(4(2) - 5) = f(8 - 5) = f(3)$

   $f(3) = 3(3) + 4 = 9 + 4 = \mathbf{13}$

e. $h(f(-3))$

   $f(x) = 3x + 4$ $h(x) = 4x - 5$

   $h(f(-3)) = h(3(-3) + 4) = h(-9 + 4) = h(-5)$

   $h(-5) = 4(-5) - 5 = -20 - 5 = \mathbf{-25}$

The absolute value function is described by the equation y = |x|. If you graph this function using a table of values as a guide, it looks something like the following two examples.

Example 2:   y = |x|

x	0	1	−1	2	−2	3	−3	4	−4
y	0	1	1	2	2	3	3	4	4

It's graph looks like this: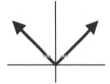

When using the vertical line test, you can see **it is a function**.

Example 3:   x = |y|

y	0	1	−1	2	−2	3	−3	4	−4
x	0	1	1	2	2	3	3	4	4

It's graph looks like this:

When using the vertical line test, you can see **it is NOT a function**.

Practice.

Let d(x) = 4x − 3 and e(x) = x² − 5x + 3. Evaluate each of the following.

1.   d(2) + e(3)   2.   d(3) − e(−2)

3.   e(a) + d(a)   4.   e(b) • d(b)

5.   e(d(3))   6.   d(e(x))

7.   e(2) − d(4)   8. $\dfrac{e(5)}{d(2)}$

9.   e(d(−3))   10.   [d(e(2))] − [e(0)]

11. $\dfrac{d(4)}{e(1)}$ 12.   d(4) − e(1)

13.   e(1) • d(4)   14.   e(d(e(1)))

15.   e(d(d(2)))   16.   d(10) + e(4)

Draw each graph and tell whether the graph represents a function. Use a table of values from −2 to 2 as a guide for the unknown. (Some problems may need more values to make a determination.)

17.   y = |x + 1| 18.   y = −|x|

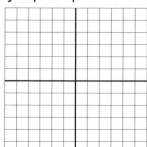   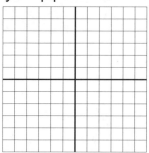

19.   x = |y + 2| 20.   y = |x + 3|

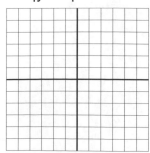   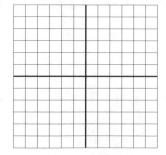

201

LESSON 69 – Graphing the Quadratic Equation: $y = ax^2$

The graph of a linear equation, $y = mx + b$ or $ax + by = c$, is a straight line. When we move to a quadratic equation such as $y = ax^2$, we have a curve.

The quadratic equation $y = ax^2$ passes through the origin since $y = 0$ when $x = 0$. We need to plot several points as shown in the following table:

x	−3	−2	−1	0	1	2	3
y	9	4	1	0	1	4	9

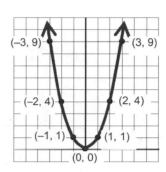

The graph of this equation is to the right. The equation $y = x^2$ defines a function and can be written $f(x) = x^2$. This graph is called a **_parabola_** with the y-axis as its axis of symmetry. As you can see from the drawing, the y-axis acts as a mirror dividing the curve into two symmetric parts. The point (0, 0) is the vertex (minimum or maximum point) of the parabola. In this example, it is the minimum point of the curve, the point for which "y" has the <u>least</u> value.

The **_domain_** (the set of replacement values allowed for "x") is any real number. The **_range_**) the set of replacement values allowed for "y") consists of all non-negative numbers.

Therefore, the domain and range for $f(x) = x^2$ is

Domain: "x" is any real number
Range: all $y \geq 0$

In summary, the steps for graphing and finding the domain and range for the quadratic equation in the form $y = ax^2$, is:

Step 1:   Set up a table of values.
Step 2:   Plot the points and draw the curve.
Step 3:   Using the graph, state the domain and range.

Example 1:   Graph $y = -x^2$ and state the domain and range.

Solution:

Step 1:

x	y
–3	–9
–2	–4
–1	–1
0	0
1	1
2	4
3	9

Step 2:

Step 3:

Domain: set of real numbers

Range: numbers that are negative or 0: $y \leq 0$

Maximum point: (0, 0)

Example 2:   Graph $y = \frac{1}{2}x^2$ and state the domain and range.

Solution:

Step 1:

x	y
–4	8
–3	9/2
–2	2
–1	12
0	0
1	1/2
2	2
3	9/2
4	8

Step 2:

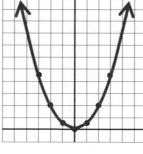

Step 3:

Domain: all real numbers

Range: all non-negative numbers

Minimum point: (0, 0)

203

Example 3: $y = 2x^2$

Step 1:

x	y
−2	8
−1	2
0	0
1	2
2	8

Step 2:

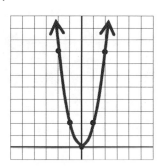

Step 3:

Domain: all real numbers

Range: $y \geq 0$

Minimum point: (0, 0)

Practice. Graph the function defined by each equation, stating its domain and range, vertex and tell whether the vertex is a minimum or a maximum point. Be sure to show a table of values for each function (values of x = −2 to 2).

1. $y = 4x^2$

x   y

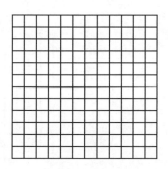

2. $y = -2x^2$

x   y

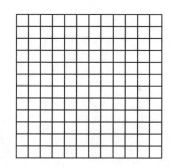

3. $y = -\frac{1}{2}x^2$

x | y

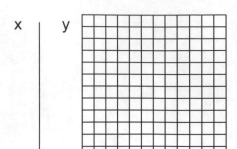

4. $y = -3x^2$

x | y

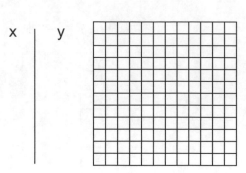

5. $y = 2x^2$

x | y

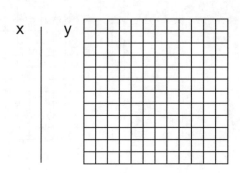

6. $y = \frac{1}{4}x^2$

x | y

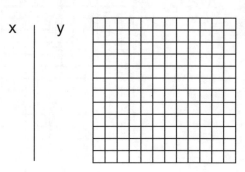

LESSON 70 – Graphing the Quadratic Equation: $y = ax^2 + k$

In the previous lesson, we graphed the quadratic equation $y = ax^2$ with the vertex at (0, 0). In the equation $y = ax^2 + k$, the vertex is not at (0, 0) but on either the x- or y-axis. For our purpose in this lesson, the vertex will lie on the y-axis.

The steps involved are:

Step 1:   Make a table of values for "x" and then solve for "y".
Step 2:   Plot the points and draw a smooth graph of the parabola curve.
Step 3:   Determine the domain, range, and vertex.

Example 1: $y = x^2 + 2$

Solution:

Step 1:

x	y
−3	11
−2	6
−1	3
0	2
1	3
2	6
3	11

Step 2:

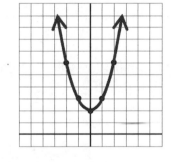

Step 3:

Domain: all real numbers

Range: $f(x) \geq 2$

Vertex: (0, 2), minimum point

Example 2: $y = -2x^2 + 5$

Solution:

Step 1:

x	y
−3	−13
−2	−3
−1	3
0	5
1	3
2	−3
3	−13

Step 2:

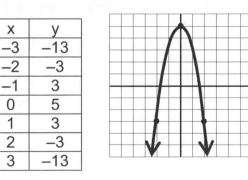

Step 3:

Domain: all real numbers

Range: $f(x) \leq 5$

Vertex: (0, 5), maximum point

Example 3: $-\frac{1}{2}x^2 - 3$

Solution:

Step 1:

x	y
−3	$-7\frac{1}{2}$
−2	−5
−1	$-3\frac{1}{2}$
0	−3
1	$-3\frac{1}{2}$
2	−5
3	$-7\frac{1}{2}$

Step 2:

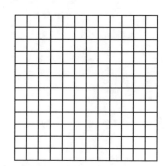

Step 3:

Domain: all real numbers

Range: f(x) ≤ −3

Vertex: (0, −3), maximum point

Practice. Graph each function defined by the equation. State its domain, range, vertex and whether the vertex is a minimum or maximum point.

1. $y = -x^2 - 2$

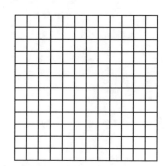

2. $y = \frac{1}{2}x^2 + 2$

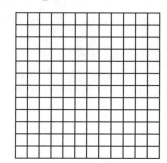

207

3. $y = -x^2 + 4$

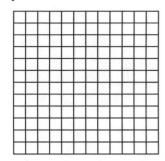

4. $y = -x^2 + 3$

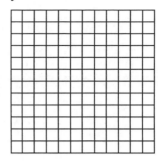

5. $y = x^2 - 1$

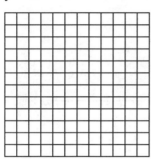

6. $y = -\frac{1}{2}x^2 - 4$

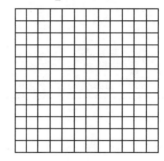

7. $y = x^2 + 1$

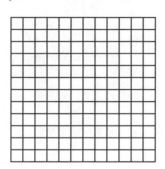

8. $y = \frac{1}{4}x^2 - 1$

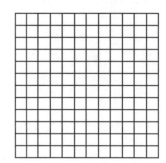

LESSON 71 – Graphing the Quadratic Equation: $y = a(x - h)^2 + k$

Previously, we graphed the quadratic equations $y = ax^2$ and $y = ax^2 + k$. We now add the quadratic equation $y = a(x - h)^2 + k$ which is graphed like the others. However, the vertex will stay on the x-axis but move left or right of (0, 0) and the axis of symmetry will be a vertical line through the vertex. Let us examine some examples.

Step 1:   Make a table of values.
Step 2:   Plot the points and draw a smooth curve.
Step 3:   Determine the domain, range, vertex and whether it is a minimum or maximum point, and the axis of symmetry.

Example 1: $y = (x - 2)^2$

Solution:

Step 1:   Step 2:   Step 3:

x	y
–1	9
0	4
1	1
2	0
3	1
4	4
5	9

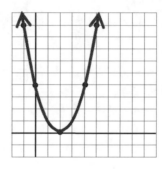

Domain: all real numbers

Range: y ≥ 2

Vertex: (3, 1), minimum point

Axis of symmetry: x = 3

Example 2: $y = (x - 3)^2 + 1$

Solution:

Step 1:   Step 2:   Step 3:

x	y
0	10
1	5
2	2
3	1
4	2
5	5

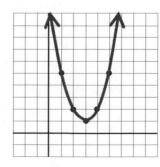

Domain: all real numbers

Range: y ≥ 0

Vertex: (2, 0), minimum point

Axis of symmetry: x = 2

Example 3: $y = -(x + 1)^2 - 2$

Solution:

Step 1:

x	y
−4	−11
−3	−6
−2	−3
−1	−2
0	−3
1	−6

Step 2:

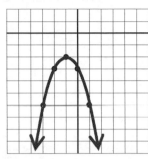

Step 3:

Domain: all real numbers

Range: $y \leq -2$

Vertex: (−1, −2), maximum point

Axis of symmetry: $x = -1$

Example 4: $y = 2(x - 2)^2 + 1$

Solution:

Step 1:

x	y
−2	33
−1	19
0	9
1	3
2	1
3	3
4	9

Step 2:

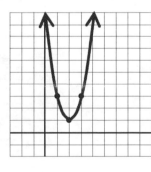

Step 3:

Domain: all real numbers

Range: $y \geq 10$

Vertex: (1, 1), minimum point

Axis of symmetry: $x = 1$

Practice. Graph each equation and identify the domain, range, vertex and maximum/minimum point, and axis of symmetry.

1. $y = (x - 1)^2 + 3$

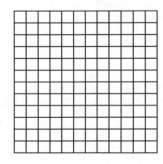

2. $y = -(x + 2)^2 + 2$

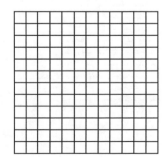

3. $y = -(x + 1)^2 - 1$

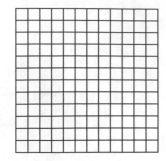

4. $y = (x - 3)^2 + 4$

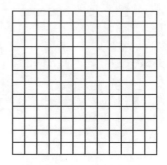

5. $y = 2(x - 1)^2 + 5$

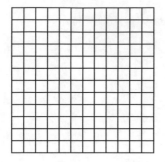

6. $y = -2(x + 1)^2 - 5$

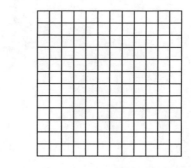

7. $y = -(x + 1)^2 - 2$

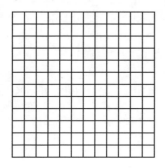

8. $y = -(x - 1)^2 - 1$

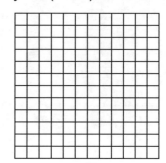

9. $y = 2(x + 3)^2 - 1$

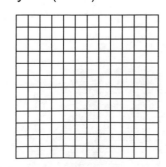

10. $y = (x + 3)^2 + 2$

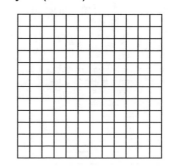

LESSON 72 – The Discriminate

Find the discriminate of any quadratic equation lets us know how many roots there are for it. The discriminate is a quantity (number) that helps us identify what type of roots the quadratic equation has. At most, any quadratic will have two roots. By finding the discriminate, we can be more precise.

Given the general form of the quadratic equation, $ax^2 + bx + c = 0$, we use the formula **$b^2 - 4ac$** to find the discriminate.

If $b^2 - 4ac > 0$, then the equation has two roots.
If $b^2 - 4ac = 0$, then the equation has one root.
If $b^2 - 4ac < 0$, then the equation has no real roots. By having **_no_** real roots, there are **_no_** x-intercepts. In other words, there are **_no_** x-values for which y = 0.

Looking at some examples, let us use the value of the discriminate to identify the number of roots for each equation.

Example 1: $x^2 - 5x - 8 = 0$

Solution: a = 1, b = –5, c = –8
$b^2 - 4ac = (-5)^2 - 4(1)(-8) = 25 + 32 = 57$
$b^2 - 4ac > 0$, therefore there are two roots.

Example 2: $x^2 + 7x + 13 = 0$

Solution: a = 1, b = 7, c = 13
$b^2 - 4ac = (7)^2 - 4(1)(13) = 49 - 52 = -3$
$b^2 - 4ac < 0$, therefore there are no real roots.

Example 3: $x^2 - 8x + 16 = 0$

Solution: a = 1, b = –8, c = 16
$b^2 - 4ac = (-8)^2 - 4(1)(16) = 64 - 64 = 0$
$b2 - 4ac = 0$, therefore there is one root.

Example 4: $3x^2 + x - 1 = 0$

Solution: a = 3, b = 1, c = –1
$b^2 - 4ac = (1)^2 - 4(3)(-1) = 1 + 12 = 13$
$b^2 - 4ac > 0$, therefore there are two roots.

Example 5: $4x^2 - 3x + 2 = 0$

Solution: $a = 4, b = -3, c = 2$
$b^2 - 4ac = (-3)^2 - 4(4)(2) = 9 - 32 = -23$
$b^2 - 4ac < 0$, therefore there are no real roots.

Example 6: $-x^2 = 2x - 3$

Solution: $x^2 + 2x - 3 = 0$ $a = 1, b = 2, c = -3$
$b^2 - 4ac = (2)^2 - 4(1)(-3) = 4 + 12 = 16$
$b^2 - 4ac > 0$, therefore there are two roots.

We can further refine the roots if the discriminate value is greater than 0. In this example, the value of the discriminate is a perfect square root, therefore both roots will be rational. Otherwise the roots will be irrational.

Practice. Use the discriminate value to tell the number of real roots of each equation and whether the roots are rational or irrational.

1. $x^2 + 6x + 9 = 0$

2. $x^2 + x + 3 = 0$

3. $4x^2 - 4x + 1 = 0$

4. $-4x^2 + 4x - 1 = 0$

5. $x^2 + 8x + 15 = 0$

6. $x^2 + 10x + 25 = 0$

7. $6x^2 + 7x - 3 = 0$

8. $2x^2 + x = 10$

9. $x^2 + 5x - 2 = 0$

10. $x^2 + 4x - 21 = 0$

11. $3x^2 + x + 2 = 0$

12. $3x^2 = 2 - x$

13. $-x^2 - x + 5 = 0$

14. $x^2 - 2x + 4 = 0$

15. $-2x^2 - x + 2 = 0$

16. $2x^2 - 3x = -1$

LESSON 73 – The Pythagorean Theorem

Having explained various quadratic equations, both in graphing and solving, we can now practice some practical applications of the quadratic.

The Pythagorean Theorem, named after its founder, lets us examine distances of the sides of a right triangle. A drawing is very helpful in understanding this mathematical concept.

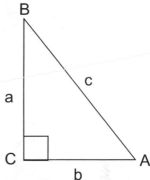

The Pythagorean Theorem is

$$a^2 + b^2 = c^2$$

where angle C ($\angle$C) is a right angle, "c" is the side opposite the right angle (called the hypotenuse), and "a" and "b" are the legs of the right triangle.

Example 1:   The hypotenuse of a right triangle is 2 inches longer than one side and 4 inches longer than the other side. Find the length of all three sides of the triangle.

Solution: Draw a triangle and label the sides and angles.

$x^2 = (x - 2)^2 + (x - 4)^2$
$x^2 = x^2 - 4x + 4 + x^2 - 8x + 16$
$x^2 = 2x^2 - 12x + 20$
$0 = x^2 - 12x + 20$
$0 = (x - 10)(x - 2)$

$x - 10 = 0$ $x - 2 = 0$
$x = 10$ $x = 2$ not a possible solution

$x - 2 = 10 - 2 = 8$
$x - 4 = 10 - 4 = 6$

Check: $(x - 2)^2 + (x - 4)^2 = x^2$
$(10 - 2)^2 + (10 - 4)^2 = (10)^2$
$8^2 + 6^2 = 10^2$
$64 + 36 = 100$
$100 \overset{\checkmark}{=} 100$

Solution: The legs of the triangle are 8 inches and 6 inches and the hypotenuse is 10 inches.

Example 2: The perimeter of a right triangle is 40 inches. If the hypotenuse is 17 inches, find the length of each of the legs.

Solution: The perimeter of the triangle minus the length of the hypotenuse is the sum of the two legs.

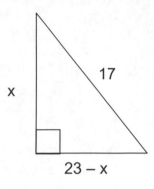

$$40 \text{ inches} - 17 \text{ inches} = 23 \text{ inches}$$

$17^2 = x^2 + (23 - x)^2$
$289 = x^2 + 529 - 46x + x^2$
$289 = 2x^2 - 46x + 529$
$0 = 2x^2 - 46x + 240$
$0 = x^2 - 23x + 120$
$0 = (x - 15)(x - 8)$

$x - 15 = 0$ $x - 8 = 0$
$x = 15$ $x = 8$

Check: $15^2 + 8^2 = 17^2$
$225 + 64 = 289$
$289 \overset{\checkmark}{=} 289$

Solution: The legs of the triangle are 15 inches and 8 inches.

Example 3: The diagonal of a rectangle is 13 feet while its perimeter is 34 feet. Find the length and width of this rectangle.

Solution: $a^2 + b2 = 13^2$ (A)
$2a + 2b = 34$ (B)

Solve equation (B) for "a".

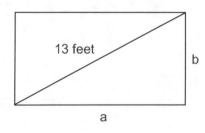

$2a + 2b = 34$
$a + b = 17$
$a = 17 - b$ (C)

Substitute equation (C) into equation (A) and solve for "b".

$(17 - b)^2 + b^2 = 13^2$
$289 - 34b + b^2 + b^2 = 169$
$2b^2 - 34b + 289 = 169$
$2b^2 - 34b + 120 = 0$
$b^2 - 17b + 60 = 0$
$(b - 12)(b - 5) = 0$

$$b - 12 = 0 \quad b - 5 = 0$$
$$b = 12 \qquad b = 5$$

Solution: When b = 12, a = 5, the length is 12 feet and the width is 5 feet. When b = 5, a = 12, the length is 5 feet and the width is 12 feet.

Practice.

1.   One leg of a right triangle exceeds the other leg by 5 inches. If the hypotenuse of the triangle is 25 inches, find the length of each leg.

2.   The hypotenuse of a right triangle is 17 inches. How long are the legs if one is 7 inches longer than the other?

3.   The perimeter of a right triangle is 36 inches and one of its legs is 3 inches longer than the other leg. Find the lengths of the three sides.

4.   The perimeter of a right triangle is 84 cm. If the hypotenuse is 35 cm, how long are the other two sides?

5.   If the leg of a right triangle is 7 inches longer than the other leg and its hypotenuse is 13 inches, find the length of both legs.

6.   The longer side of a rectangle is 4 cm greater than the other side. If the diagonal of the rectangle is 20 cm, find the length of both sides.

7.   The altitude of an isosceles triangle is 4 cm less than either equal side. The base is 16 cm in length. Find the perimeter of the isosceles triangle.

8.   A right triangle has a perimeter of 56 inches and a hypotenuse of 25 inches. What are the lengths of the legs?

9.   The diagonal of a rectangle is 25 cm. The length and width of the rectangle total 35 cm. Find the length and width.

10.   A square has a triangle attached to either side to form a trapezoid. If the side of the square is 12 inches and each hypotenuse of the triangle is 15 inches, find the area of the figure.

LESSON 74 – The Distance Formula

The distance between any two points (x_1, y_1) and (x_2, y_2) is given by the formula:

$$d = \sqrt{(x_2 - x_1)^2 + (y_2 - y_1)^2}$$

Example 1:   Find the distance between $(-2, 1)$ and $(3, 4)$.

   Solution: Let $(x_1, y_1) = (-2, 1)$ and $(x_2, y_2) = (3, 4)$.

$$d = \sqrt{(x_2 - x_1)^2 + (y_2 - y_1)^2}$$
$$d = \sqrt{(3 - (-2))^2 + (4 - 1)^2}$$
$$d = \sqrt{5^2 + 3^2}$$
$$d = \sqrt{25 + 9}$$
$$d = \sqrt{34}$$

Example 2:   Find the distance between $(3, -6)$ and $(2, 7)$.

   Solution: $d = \sqrt{(x_2 - x_1)^2 + (y_2 - y_1)^2}$
$$d = \sqrt{(2 - 3)^2 + (7 - (-6))^2}$$
$$d = \sqrt{(-1)^2 + 13^2}$$
$$d = \sqrt{1 + 169}$$
$$d = \sqrt{170}$$

Practice.

1.   Find the distance between the two points. Express irrational answers in simplest radical form.

   a. $(-7, 0)$ and $(0, 5)$ b. $(-3, 4)$ and $(6, -5)$

   c. $(6, 2)$ and $(-4, 3)$ d. $(2, -3)$ and $(-4, 7)$

   e. $(10, 4)$ and $(3, -6)$ f. $(-4, 2)$ and $(5, 8)$

   g. $(8, -6)$ and $(-3, 7)$ h. $(0, 3)$ and $(7, 0)$

2. Find the perimeter of the triangle with vertices at $A(-3, 1)$, $B(2, 5)$, and $C(4, -4)$.

3. Find the perimeter of the quadrilateral shown with vertices at A(–2, 2), B(2, 6), C(5, 3), and D(1, –1).

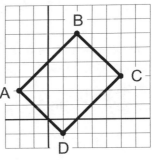

4. Three vertices of a rectangle are E(–3, 5), F(1, 5) and g(–3, –4). Find the coordinates of the fourth vertex, H. Then find the length of the diagonal, its perimeter and its area.

5. Find the radius of a circle with center at (1, 2) passing through the point (–3, 7).

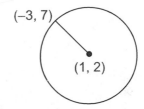

6. A diameter of a circle has endpoints at (–2, –2) and (6, 2). Find the length of a radius of the circle.

7. Three vertices of a square are A(–3, 5), B(–1, 1) and C(3, 3). Find the coordinates of the fourth vertex, D. Then find the length of a diagonal and the area of the square.

8. Find the perimeter of a quadrilateral with vertices at W(10, –3), X(2, 1), Y(1, –1) and Z(5, –3).

9. A triangle has vertices at D(6, 3), E(4, –3) and F(2, 1). Find the area of the triangle.

10. Given a right triangle with vertices at M(–4, –5), N(–4, 7), and P(4, 7), find the area and perimeter of the triangle.

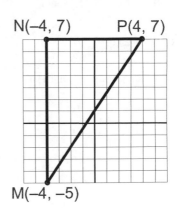

LESSON 75 – Radical Equations

Radical equations contain a variable under the radical sign ($\sqrt{}$). We can solve these equations by several techniques. Here are some examples to practice and understand.

Example 1:   Solve $\sqrt{x} = 5$.

Solution: Square both sides.

$$(\sqrt{x})^2 = 5^2$$
$$x = 25$$

Check: $\sqrt{25} = 5$
$5 \overset{\checkmark}{=} 5$

Solution: $x = 25$

Example 2:   Solve $\sqrt{x + 3} = 8$.

Solution: Square both sides.

$$(\sqrt{x + 3})^2 = 8^2$$
$$x + 3 = 64$$
$$x = 61$$

Check: $\sqrt{61 + 3} = 8$
$\sqrt{64} = 8$
$8 \overset{\checkmark}{=} 8$

Solution: $x = 61$

Example 3:   Solve $\sqrt{a} + 5 = 12$

Solution: Isolate the radical by subtracting 5 from both sides.

$$\sqrt{a} = 7$$

Square both sides.

$$(\sqrt{a})^2 = 7^2$$
$$a = 49$$

219

Check: $\sqrt{49} + 5 = 12$

$7 + 5 = 12$

$12 \overset{\checkmark}{=} 12$

Solution: a = 49

Example 4: Solve $\sqrt{b} + 6 = 2$

Solution: Isolate the radical by subtracting 6 from both sides.

$\sqrt{b} = -4$

Square both sides.

$(\sqrt{b})^2 = (-4)^2$
$b = 16$

Check: $\sqrt{16} + 6 = 2$
$4 + 6 = 2$
$10 \neq 2$

Solution: Since $10 \neq 2$, there is **no solution** to this problem!!

BE CAREFUL!!! SOMETIMES RADICAL EQUATION <u>DO NOT</u> HAVE A SOLUTION. MAKE CERTAIN YOU CHECK YOUR WORK THOROUGHLY!!!

Practice. Solve and check.

1. $\sqrt{x - 2} = 8$

2. $\sqrt{a} = 7$

3. $6 = \sqrt{b}$

4. $\sqrt{3x} = 6$

5. $2\sqrt{y} = 4$

6. $\sqrt{3 + x} = -8$

7. $\sqrt{3x - 8} = 2$

8. $2\sqrt{a} + 3 = 9$

9. $\sqrt{x} + 4 = 7$

10. $2\sqrt{x+1} = 8$

11. $3\sqrt{x} + 9 = 6$

12. $\sqrt{a+5} = 4$

13. $16 = \sqrt{x-2} + 3$

14. $\dfrac{\sqrt{x}}{2} = 6$

15. $3\sqrt{\dfrac{x}{4}} = 2$

16. $\dfrac{\sqrt{a+4}}{3} = 7$

17. $\sqrt{x^2} = -5$

18. $\sqrt{3x} = 4\sqrt{3}$

19. $\sqrt{a+4} = 5\sqrt{2}$

20. $\sqrt{y^2} + 5 = 12$

21. $\sqrt{y+3} + y = 3$

22. $\sqrt{2a-2} = \sqrt{a+1}$

23. $\sqrt{x^2} + 1 = 1$

24. $\sqrt{x+5} - 3 = x$

25. $\sqrt{b+12} - b = 10$

LESSON 76 – Direct and Inverse Variation

Our last lesson for Algebra I discusses the direct and inverse variation. They differ in that the direct variation is the **ratio** between two numbers, is always the same, and inverse variation, the **product** of the two numbers is always the same.

We shall examine direct variation with a few examples first.

Example 1:   If "y" varies directly as "x", and y = 20 when x = 30, find (a) the constant "k" and (b) find "y" when x = 45.

   Solution: Direct variation is a **ratio**.

   Step 1:   Find "k".

   $$k = \frac{y}{x}$$
   $$k = \frac{20}{30}$$
   $$k = \frac{2}{3}$$

   Step 2:   Find "y".

   $$y = kx$$
   $$y = \frac{2}{3} \cdot 45$$
   $$y = 30$$

   Solution: (a) $k = \frac{2}{3}$ and (b) y = 30

Example 2:   The shadows of two buildings vary directly as their heights at any given time. If a building 80 feet tall casts a 25-foot shadow. how tall is a building that casts a 15-foot shadow?

   Solution: This is solved by using proportions.

   $$\frac{\text{height of building 1}}{\text{shadow of building 1}} = \frac{\text{height of building 2}}{\text{shadow of building 2}}$$

222

$$\frac{80 \text{ feet}}{25 \text{ feet}} = \frac{h \text{ feet}}{15 \text{ feet}} \quad \text{cross multiply}$$

$$(25)(h) = (80)(15)$$
$$25h = 1200$$
$$h = 48$$

Solution: The height of the building is 48 feet.

Example 3:   If "y" varies directly as "x" and x = 4 when y = 12, then what is "x" when y = 21?

Solution: You can use either method.

$$\frac{y}{x} = k \qquad\qquad \frac{y_1}{x_1} = \frac{y_2}{x_2}$$

$$\frac{12}{4} = k \qquad\qquad \frac{12}{4} = \frac{21}{x}$$

$$3 = k \qquad\qquad 12x = 4(21)$$
$$\qquad\qquad\qquad 12x = 84$$

$$\frac{21}{x} = 3 \qquad\qquad x = 7$$
$$3x = 21$$
$$x = 7$$

Solution: x = 7 when y = 21

Either method can be used to solve direction variations.

Now for some inverse variation examples.

Example 4:   Give a formula to show how "b" varies inversely with "e", using "k" as the constant.

Solution: Inverse variation is a **product**. Remember, their products equal the constant.

$$eb = k$$

Example 5:   Suppose "y" varies inversely as "x", and x = 2 when y = 1. Find (a) the constant and (b) when x = 4, the "y" is what?

Solution: (a) $xy = k$
                        $2 \cdot 1 = k$
                        $2 = k$

                  (b) $xy = k$
                        $4y = 2$
                        $y = \frac{1}{2}$

Solution: (a) $k = 2$ and (b) $y = \frac{1}{2}$

Example 6:   Electrical resistance (R) in a wire varies **directly** as the length (L) and **inversely** as the square of the diameter (D). (a) Express this variation in equation form. (b) If 4250 ft of a $\frac{5}{32}$ -inch wire has a resistance of 13.6 ohms, how long would a $\frac{3}{16}$ -inch wire have to be in order to have a resistance of 4.2 ohms?

Solution: (a) $R = \frac{kL}{d^2}$

                  (b) $13.6 = \dfrac{4250k}{\left(\frac{5}{32}\right)^2}$
                        $k = 7.8125 \times 10^{-5}$
                        $4.2 = \dfrac{L(7.8125 \times 10^{-5})}{\left(\frac{3}{16}\right)^2}$
                        $1890 = L$

Solution: (a) $R = \frac{kL}{d^2}$ and (b) $k = 7,8125 \times 10^{-5}$ and L = 1890 feet

Practice.

1. If "y" varies directly as "x", and y = 16 when x = 12, find

   a. "y" when x = 48   b. "y" when x = 21
   c. "x" when y = 40   d. "x" when y = 8
   e. "x" when y = 48   f. "y" when x = 24
   g. "x" when y = 2   h. "y" when x = 60

2. If "y" varies inversely as "x", and x = 40 when y = 8, find

   a. "x" when y = 16   b. "y" when x = 10
   c. "x" when y = 12   d. "y" when x = 20
   e. "y" when x = 64   f. "y" when x = 320
   g. "x" when y = $\frac{5}{2}$ h. "x" when y = $\frac{1}{2}$

3. Identify each equation as representing direct or inverse variation.

   a. P = 1.2C   b. $y = -\frac{5}{x}$ c. ab = 20   d. D = πd

4. "N" varies inversely as "y" and N = 20 when y = 0.35. Find "y" when N = 2.

5. If "y" varies inversely as the square of "x" and y = 12 when x = 2, find "y" when x = 6.

6. Express as an equation: The area of a triangle varies jointly as its base, "b", and its altitude, "h". What is "k" in your answer?

7. Express as an equation: The perimeter, "p", of a hexagon, varies directly as the length of a side, "s". What is the constant (k) value in your equation?

8. Find the missing numbers in the following table knowing "h:d" is constant.

h	1	2	4	b	7	d
d	40	80	a	200	c	360

9. By means of a formula, express the relation between the variables knowing they vary inversely.

a.

x	1	2	4	8
y	40	20	10	5

b.

R	2	4	10	20
T	50	25	10	5

c.

C	12	24	36	48
D	12	6	4	3

A 75-Minute Review
of
Fundamentals of Math Book 1

This is a review for students prior to taking Algebra 1. If you score 40 or more correct, proceed to the beginning of this book. If less than 40, it might be advisable to go back and review my pre-algebra book which can be purchased for less than $10 at:

www.ortner-algebra.com

The ISBN for Fundamentals of Math Book 1 is: 978-1-4343-5875-2

Numbers in grey refer to the Lessons in Fundamentals of Math Book 1.

1.   (25) $2\frac{1}{3} + 3\frac{4}{5} + 6\frac{3}{4}$ leave answer as mixed number in lowest terms

2.   (31) $\dfrac{x}{20} = \dfrac{6}{40}$

3.   (21) $\frac{5}{6} \times \frac{3}{10}$

4.   (8)   Change 21.68 to a mixed number in lowest terms.

5.   (6) $27.23 \cdot 10.4$

6.   (22) $\dfrac{\frac{3}{4}}{\frac{7}{16}}$

7.   (40)   Sally bought 6 pens for $14.04. Each pen cost her how much?

8.   (45)   What is the 5% sales tax on a $122 radio?

9.   (52)   Solve for x: $x - 32 = 78$

10.   (54)   Simplify: $6^2 + 4 - 3$

11.   (43)   Sam ate $\frac{2}{3}$ of the box of cookies. There were 24 cookies in the box. How many cookies did Sam eat?

12.   (38)   What percent of 30 is 12?

13.   (29)   Express $\frac{78}{100}$ as a percent.

14. (28) $(6\frac{1}{2} + 5\frac{2}{3}) - 4\frac{2}{9}$ leave answer as a mixed number in lowest terms.

15. (55) Simplify: $5^2 + 3(6 - 2) - 18$

16. (62) Solve for x: $0.5x + 0.7x = 40 - 2^2$

17. (69) $-3^2 + (-5)^2$

18. (70) solve for x: $3x - 5 = x + 7$

19. (23) $5\frac{3}{5} \cdot 4\frac{1}{9}$ leave answer as a mixed number in lowest terms

20. (14) Given $17y^3$, name the exponent, variable, and coefficient.

21. (60) Simplify: $-[-(8)] + [-2 + (-3) - 5]$

22. (64) Simplify: $-3(2a - 3b + 4c - 5d)$

23. (34) $2\frac{1}{2}$ of what number is 30?

24. (36) 80% of what number is 400?

25. (44) If 10 pounds of sugar cost $16, what will 25 pounds of sugar cost?

26. (15) Find the Greatest Common Factor of 48 and 60.

27. (50) Simplify: $16 + 4 \cdot (-5) - 8$

28. (57) $\sqrt[3]{125}$ equals _____

29. (66) Simplify: $\dfrac{(-3)(+4) + (8)(7)}{(-7) + (-4)}$

30. (68) Simplify: $4x + 5y - 3z + 5x - 8y - 10z$

31. (13) What is the least common multiple of 6, 24, and 30?

32. (11) What is the prime factorization of 72?

33. (65) Find "x", given $3x - 5y = 4 + 6y$ if $y = 4$.

34. (48) Simplify: $(-3) + (-5) + 9 + (-6)$

35. (27) $(6\frac{2}{3} + 4\frac{1}{4}) - (3\frac{1}{8} - 2\frac{4}{5})$

36. (67) Simplify: $\dfrac{(-14)(12)+(-2)(3)}{2 + (-8)}$

37. (42) A car travels 720 miles in 12 hours. What is the average speed of the car?

38. (35) Three-fifths of what number is 69?

39. (56) Evaluate: $2\sqrt{5}(\sqrt{5}) - \sqrt{49}$

40. (18) $\frac{3}{8} \div \frac{7}{20}$

41. (33) Change $\frac{7}{20}$ to a decimal and a percent.

42. (58) Evaluate: $(-3)^2 + \sqrt[3]{64} - (-2)^3$

43. (16) Reduce $\frac{36}{60}$ to lowest terms.

44. (63) Solve for "x": $14\frac{2}{3} - x = 6\frac{7}{10}$

45. (61) Solve for "x": $\frac{2}{3}x + \frac{5}{8}x = 2\frac{5}{12} + 6\frac{7}{16}$

46. (71) Solve for "x": $2x - 8 + 6x + 16 = 4x + 32$

47. (51) Evaluate: $-(a + b^2) + a^b - 2c$ where a = 4, b = 2, and c = 6

48. (41) If a car travels 62 miles on 2 gallons of gas, how many gallons does it need to travel 155 miles?

49. (47) Change $8\frac{1}{2}\%$ to a fraction and a decimal.

50. (39) Solve the proportion and simplify your answer if possible: $\dfrac{3\frac{2}{3}}{x} = \dfrac{5\frac{1}{6}}{20\frac{3}{4}}$

ANSWERS

LESSON 1

1. 2	15. -10	27. (c) 4	31. (a) -1280	34. (c) 4	39. b
2. -20	16. -14	(d) -5	(b) 300	(d) -5	40. c
3. -12	17. 2	28. (a) 42	(c) 8	35. (a) 24	41. a
4. 2	18. -3	(b) -28	(d) -20	(b) -16	42. d
5. 5	19. 7	(c) 5	32. (a) 72	(c) 3	43. c
6. -10	20. -18	(d) -4	(b) -24	(d) -4	44. c
7. 5	21. -13	29. (a) -1120	(c) 4	36. (a) -45	45. a
8. -8	22. -6	(b) 550	(d) -3	(b) -52	46. d
9. -1	23. 10	(c) 10	33. (a) -640	(c) 144	47. b
10. 10	24. 2	(d) -50	(b) 850	(d) 96	48. b
11. 10	25. -12	30. (a) -570	(c) 9	37. b	49. d
12. 2	26. 7	(b) 980	(d) -30	38. a	50. b
13. -3	27. (a) 4	(c) 9	34. (a) 20		
14. 1	(b) -27	(d) -70	(b) -10		

LESSON 2

1. a	11. d	21. b	31. 31	40. b	49. 50	58. 171
2. 15	12. b	22. d	32. 14	41. 42	50. 13	59. 20
3. 9	13. c	23. -4	33. d	42. 148	51. 10	60. 11
4. 9	14. d	24. a	34. b	43. b	52. 16	61. 45
5. a	15. d	25. a	35. 22	44. 114	53. 0	62. 8
6. b	16. d	26. 2	36. 14	45. b	54. 25	63. -28
7. c	17. -16	27. -148	37. -18	46. 22	55. 78	64. 24
8. d	18. 6	28. -19	38. 552	47. 31	56. 44	65. 3
9. c	19. -208	29. b	39. 86	48. 22	57. 20	66. 18
10. 19	20. 61	30. 28				

LESSON 3

1. b	5. b	9. -99	13. 15/11	17. 4	21. 2	24. 20
2. b	6. d	10. 427	14. -13	18. 6	22. -24	25. 160
3. 18	7. 77	11. 3	15. 88	19. 34	23. -19	26. 10
4. 34	8. -26	12. 3/17	16. 21	20. 36		

LESSON 4

1. 3	10. 13	19. 4	28. 19	37. 20	45. 16	53. d
2. 6	11. 24	20. 10	29. 8	38. 42	46. d	54. 2
3. 7	12. 8	21. 2	30. 6	39. 15	47. b	55. 3
4. 12	13. 21	22. 21	31. 21	40. -2	48. c	56. -3
5. 10	14. 14	23. 15	32. 30	41. 10	49. 4	57. -17
6. 9	15. 24	24. 72	33. 18	42. -8	50. b	58. 23
7. 11	16. 4	25. 80	34. 12	43. 16	51. a	59. b
8. 6	17. 7	26. 3	35. 25	44. -3	52. b	60. -1
9. 15	18. 35	27. 5	36. 60			

LESSON 5

1. 3	6. 2	11. 11	16. 16	21. d	26. 23	30. c
2. -20	7. 3	12. -26	17. 0	22. d	27. -1	31. a
3. 16	8. 8	13. d	18. -13	23. -14	28. -1	32. b
4. 23	9. 12	14. c	19. -16	24. -20	29. c	33. 16
5. -4	10. b	15. -10	20. c	25. -8		

LESSON 6

1. -27	**5.** -39	**9.** -24	**13.** -3	**17.** -13	**21.** -52	**25.** -8
2. -16	**6.** -49	**10.** -7	**14.** 11	**18.** -25	**22.** -43	**26.** -20
3. -9	**7.** 6	**11.** 20	**15.** -26	**19.** -24	**23.** 55	**27.** -20
4. 19	**8.** 30	**12.** -31	**16.** 21	**20.** 10	**24.** -23	**28.** -27

LESSON 7

1. N = 26	**6.** r = 24	**11.** g = 25	**16.** f = 8	**21.** j = 22	**26.** p = -24	**30.** N = -94
2. N = 34	**7.** t = 16	**12.** j = 29	**17.** e = 6	**22.** d = 20	**27.** Q = -25	**31.** S = -95
3. n = 31	**8.** c = 14	**13.** N = 27	**18.** N = 37	**23.** N = 27	**28.** M = 30	**32.** S = 113
4. y = 11	**9.** N = 28	**14.** N = 31	**19.** c = 28	**24.** N = 37	**29.** a = 38	**33.** Z = 63
5. j = 36	**10.** N = 26	**15.** N = 33	**20.** v = 16	**25.** x = -42		

LESSON 8

1. h = 8	**6.** s = 8	**11.** w = 3	**15.** n = 60	**19.** N = 15	**23.** N = 15	**27.** f = 80
2. s = 2	**7.** z = 40	**12.** f = 4	**16.** c = 30	**20.** j = 80	**24.** u = 21	**28.** c = 12
3. m = 7	**8.** m = 5	**13.** g = 18	**17.** N = 35	**21.** t = 12	**25.** d = 33	**29.** u = 40
4. c = 3	**9.** b = 3	**14.** m = 6	**18.** N = 42	**22.** x = 99	**26.** p = 18	**30.** x = 21
5. p = 15	**10.** r = 9					

LESSON 9

1. x = 2	**4.** x = 5	**6.** x = 5/2	**8.** x = -8/9	**10.** x = 6	**12.** x = 10	**14.** x = 7
2. x = 5	**5.** x = 1/3	**7.** x = -2/3	**9.** x = 2	**11.** x = 7	**13.** x = 2	**15.** x = 1
3. x = 4						

LESSON 10

1. x = 8/7	**6.** x = 1	**11.** x = -1/3	**16.** x = -1	**21.** x = -5	**25.** x = 6
2. x = -1	**7.** x = 1	**12.** x = 1/4	**17.** x = 2	**22.** x = -4	**26.** x = 4
3. x = -2/3	**8.** x = 8/7	**13.** x = 10	**18.** x = -7/18	**23.** x = 2	**27.** x = 6
4. x = -5	**9.** x = -3	**14.** x = -3	**19.** x = -10/17	**24.** x = 18	**28.** x = -6
5. x = -10/3	**10.** x = -1/6	**15.** x = -5	**20.** x = -6		

LESSON 11

1. x = -1	**6.** x = 4/9	**11.** x = 1/8	**16.** x = 4/3	**21.** x = 2	**25.** x = 2/3
2. x = 2	**7.** x = -6	**12.** x = -9/5	**17.** x = -1/7	**22.** x = 1	**26.** x = 5
3. x = -7	**8.** x = 3	**13.** x = -4/5	**18.** x = -6	**23.** x = 5	**27.** x = 4
4. x = 3	**9.** x = 1	**14.** x = 6	**19.** x = 4	**24.** x = -6	**28.** x = 8
5. x = 5	**10.** x = -2/7	**15.** x = -2	**20.** x = -2		

LESSON 12

1. x = -4	**4.** x = 10	**7.** x = 10	**10.** x = -0.2	**12.** x = 0.65	**14.** x = -3
2. x = -4	**5.** x = 1.5	**8.** x = 8	**11.** x = -3	**13.** x = -3	**15.** x = -2
3. x = -5	**6.** x = 6	**9.** x = -5.2			

LESSON 13

1. x = 21	**6.** x = -35	**11.** a = 5/4	**16.** x = 257/24	**21.** x = -1/4	**26.** x = -1/13
2. y = -3/2	**7.** x = 31/12	**12.** x = 1/2	**17.** y = 9/2	**22.** x = -45/38	**27.** x = 7/22
3. x = 10	**8.** x = 43/6	**13.** y = 4/15	**18.** x = 102/7	**23.** x = 13/20	**28.** x = 17/21
4. x = -6	**9.** x = 6	**14.** x = -3/7	**19.** x = 2/3	**24.** x = 5	**29.** x = 5/2
5. x = -10	**10.** x = -31/3	**15.** y = 11/9	**20.** x = -17/9	**25.** x = -15/4	**30.** x = 296/27

LESSON 14

1. h = -0.9
2. g = -0.8
3. k = 28
4. s = 23
5. g = 10

6. w = 18
7. c = 5
8. s = 3
9. h = 2
10. x = 5.6

11. w = 4.8
12. h = 54
13. y = 49
14. p = 35
15. y = 18

16. y = 12
17. y = 12
18. n = -0.7
19. a = 8
20. u = 5

21. w = -0.7
22. t = -0.6
23. b = 353/13
　　 = 27 2/13
24. p = 1.5

25. N = 15
26. x = -40
27. A = 2
28. B = 48

LESSON 15

1. $r = \dfrac{E}{a}$

2. $W = \dfrac{A}{L}$

3. $t = \dfrac{i}{pr}$

4. $a = \dfrac{2S}{t^2}$

5. b = S − a − c

6. $W = \dfrac{P - 2L}{2}$
　　 or $W = \dfrac{P}{2} - L$

7. $t = \dfrac{A - p}{pr}$

8. $q = \dfrac{D - r}{d}$

9. $A = \dfrac{90S}{\pi r^2}$

10. $g = \dfrac{-Rs}{R - s}$
　　 or $g = \dfrac{Rs}{s - R}$

11. $f = \dfrac{mg - T}{m}$
　　 or $f = g - \dfrac{T}{m}$

12. $R = \dfrac{E - Ir}{I}$
　　 or $R = \dfrac{E}{I} - r$

13. $q = \dfrac{fp}{p - f}$

14. r = D − qd

15. k = a − 6mq

16. $r = \dfrac{a - p}{pt}$
　　 or $r = \dfrac{a}{pt} - \dfrac{1}{t}$

17. $R = \dfrac{P(1 - TO)}{Q}$

18. $W = 3H - \dfrac{9}{2}$
　　 or W = 3H − 4.5

19. $x = \dfrac{z}{y} - 2$
　　 or $x = \dfrac{z - 2y}{y}$

20. $b = \dfrac{d}{ac} - d$
　　 or $b = \dfrac{d - acd}{ac}$

LESSON 16

1. 12
2. 30
3. x = 11
4. 2
5. d
6. x = 6

7. $h = \dfrac{2A}{b}$
8. -5x + 2y − 6z
9. x = 20.275
10. 2a + 2b + 48
11. c

12. x = 247/20
　　 or x = 12.35
13. -2x + 7
　　 or 7 − 2x
14. 3
15. true

16. a = 17/5
　　 or x = 3.4
17. 15
18. $r = \dfrac{C}{2\pi}$
19. x = 325/42
20. x = 10

LESSON 17

1. a
2. $25,000,000
3. 2001
4. $10,000,000

5. (a) 2003
　　 (b) 2002
6. (a) 2003
　　 (b) 12%
7. 20 students

8. 154 students
9. 45 students
10. 15 students
11. c
12. 100 students

LESSON 18

1. (a) yes
　　 (b) no
　　 (c) yes
　　 (d) no
2. (a) no
　　 (b) no
　　 (c) yes
　　 (d) yes
3. (a) no
　　 (b) yes
　　 (c) yes
　　 (d) yes

4. (a) yes
　　 (b) no
　　 (c) no
　　 (d) no
5. y = -5
6. y = -4
7. x = 6
8. y = 6
9. x = 5
10. y = -4
11. x = -1

12. (a) y = -7
　　 (b) x = 6
　　 (c) x = 2
　　 (d) y = 13
13. (a) y = 5
　　 (b) x = 4
　　 (c) y = -4
　　 (d) x = 0
14. (a) y = 15
　　 (b) y = 5
　　 (c) x = 5
　　 (d) x = 8

15. (a) y = 11
　　 (b) x = -4
　　 (c) y = 20
　　 (d) x = -6
16. (a) x = 4
　　 (b) y = -6
　　 (c) y = -2
　　 (d) x = -6

17. (a) y = 3
　　 (b) x = 7/4
　　 (c) y = 1
　　 (d) x = -3/4
18. (a) x = 15/2
　　 (b) y = -8/3
　　 (c) x = 6
　　 (d) y = -6

LESSON 19

1.

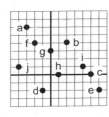

2. (a) (2,4)
   (b) (-7,2)
   (c) (2,-7)
   (d) (-5,-4)
   (e) (0,7)
   (f) (-3, 0)
   (g) (8,-5)
   (h) (5,9)
   (i) (-2,-5)
   (j) (4,0)

3. (a) (6,1)
   (d) (0,-5)
   (e) (-2,6)
   (g) (5,-2)
   (i) (-6,0)

LESSON 20

1. intersect   **2.** parallel   **3.** intersect   **4.** $y = -\frac{2}{3}x$

LESSON 21

1. $m = \frac{8-7}{5-3} = \frac{1}{2}$

2. $m = \frac{9-(-3)}{6-7} = \frac{9+3}{-1} = \frac{12}{-1} = -12$

3. $m = \frac{15-6}{8-(-2)} = \frac{9}{8+2} = \frac{9}{10}$

4. $m = \frac{-9-(-6)}{-7-(-4)} = \frac{-9+6}{7+4} = \frac{-3}{-3} = 1$

5. $m = \frac{11-5}{8-5} = \frac{6}{3} = 2$

6. 1/2

7. 5/3   **10.** 0   **13.** -5

8. 2   **11.** -1   **14.** -4/3

9. 1   **12.** 2/5

LESSON 22

1. (0,3), (9,0)   **4.** (0,-6), (-3,0)   **7.** (0, 17/8), (-17/4,0)   **10.** (0,-3), (3,0)   **13.** (0,3/4), (-3,0)
2. (0,-10/3), (5,0)   **5.** (0,10), (5/2,0)   **8.** (0,-4), (5,0)   **11.** (0,1/2), (-1/4,0)   **14.** (0,0.5), (-0.3,0)
3. (0,3), (-4,0)   **6.** (0,5), (-3,0)   **9.** (0,2), (2,0)   **12.** (0,5/3), (5/2,0)

LESSON 23

1. x = 2   **3.** y = -5   **5.** y = 0   **7.** y = -10   **9.** y = 3/4
2. y = 4   **4.** x = -3   **6.** x = 2/3   **8.** x = 7   **10.** y = -8

LESSON 24

1. m = -3
   zeros: (0,0)

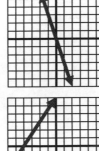

2. m = 2/3
   zeros: (0,-6), (9,0)

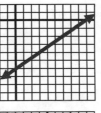

3. m = 3/2
   zeros: (0,6), (-4,0)

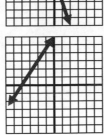

4. m = undefined
   zeros: (3,0)
   parallel to y-axis

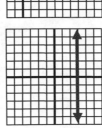

5. m = 2/3
   zeros: (0,-3/2),
        (9/4,0)

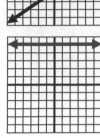

6. m = 1/2
   zeros: (0,-1), (2,0)

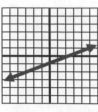

7. m = 0
   zeros: (0,5)
   parallel to x-axis

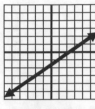

8. m = -2/3
   zeros: (0,5), (15/2,0)

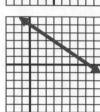

LESSON 25

1. slopes are the same, parallel lines
2. product of slopes = -1, perpendicular lines
3. slopes are the same, parallel lines
4. slopes different, neither
5. slopes different, neither

6. product of slopes = -1, perpendicular lines
7. product of slopes = -1, perpendicular lines
8. slopes different, neither
9. slopes different, neither
10. product of slopes = -1, perpendicular lines

LESSON 26

1. $3x - y = -1$ **2.** $3x - 4y = -16$ **3.** $x - 2y = -4$ **4.** $22x + 18y = 17$ **5.** $3x + 9y = 22$

LESSON 27

1. $2x - y = -2$ **3.** $2x + y = 0$ **5.** $5x + 6y = -4$ **6.** $4x + 5y = 22$ **7.** $5x + 2y = -1$
2. $y = -8$ **4.** $2x + y = 18$

LESSON 28

1. zeros: (0,3), (-3/2,0)
   m = 2
   ≤ means solid line
   (0,0) in shaded region
   selected point: (4,-5)

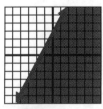

2. zeros: (0,4), (2,0)
   m = -2
   > means dashed line
   (0,0) not in shaded region
   selected point: (6,4)

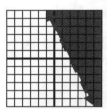

3. zeros: (0,-1),(-1/3,0)
   m = -3
   < means dashed line
   (0,0) not in shaded
      region
   selected point: (-3,-1)

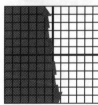

4. zeros: (0,-5), (5,0)
   m = 1
   ≥ means solid line
   (0,0) is in the shaded
      region
   selected point: (-3,-4)

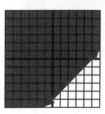

LESSON 29

1. no, slopes different
2. $y = \frac{3}{4}x + 5$ or $3x - 4y = -20$
3. no
4. $x = 6$
5. m = 3/2
6. (-1,2), (5,0)
7. m = 3

8. (0,8/5), (4,0)
9. $y = 3x - 2$
10. horizontal
11. yes
12. $5x + 2y = 0$
13. yes
14. $3x - y = 22$

15. no
16. m = -3
17. (0,14/5), (14/3,0)
18. true
19. false
20. (a) 35%
   (b) 42 people

LESSON 30

1. $30g^2h^5 - 12g^6h$
2. $8b^4c^5 + 12b^7c^2$

3. d

4. (a) 4
   (b) $1/4$
   (c) 1
5. a
6. (a) x^8y^6z
   (b) $\dfrac{x^8y^{10}}{z}$
7. (a) $x^{-1}y^6z$
   (b) $x^{-1}y^4z$
   (c) $3x^{-9}y^{-6}z^{25}$
8. $ax + 2ay$
9. $14 + 21x$

10. $-3y^3 - 9yz$
11. $2x^4 + 6x^3 + 8x^2$
12. $-5cd^2 - 5cef$
13. $8x^2 + 12x$
14. $-20a - 40b$
15. $40a^3 + 48a^2 + 32a$
16. $-m - n - 2p$
17. $30 + 90x$
18. $6m^3 + 18m^2 + 36m$
19. $-9e^2f - 9e^2g$
20. $-7x^2 - 7xy$
21. $12x^3 + 8x^2 + 16x$
22. $6x + 8y + 10z$

23. $-5t^4 - 15t^2 - 20t$
24. $12xy^2 + 28x^2yz$
25. $\dfrac{12a}{b^2} - \dfrac{21a^3}{c^2}$
26. $12x^3 - 15x^2 + 18x$
27. $36x^2 - 6x + 1$
28. $-x^5 + 4x^4 + 3x^3 - 7x^2$
29. $8x^2 + 3x - 20$
30. $6x^3 + 24x^2 + 36x$
31. $100y^3 - 60y^2 - 15$
32. $-a^3 + a^2b + 5ab^2$
33. $-32e^4 + 16e^3 - 8e^2 - 24e$
34. $abc^2 - 3abc + 4ab$
35. $-9f^3 + 36f^2 - 3f$

LESSON 31

1. $14x^3 + 12x - 12$
2. $-2x^3 + 12x^2 - 3x + 4$
3. $4a^5 - 12a^2 - 9a + 2$
4. $-8y^4 + 14y^2 - 3y + 30$
5. $-48x^5 - 38x^3 + 5x - 20$

6. $6x^5 + 13x^4 + 16x^3 + 27x^2 + 16x - 11$
7. $-16x^5 + 4x^4 - 14x^3 + 24x^2 + 34x - 16$
8. $-9x^2 - 6x + 3$
9. $4z^2 - 1$
10. $15a^4 - 20a^3 + 18a^2 + 2a - 3$

11. $2x^2 - 12x - 6$
12. $7n^2 - 4n - 4$
13. $4a^3 - 8a^2 + 14a - 10$
14. $-6y^3 + 12y^2 - 9y + 1$

LESSON 32

1. $-4m^2 - 9mn - 2n^2$
2. $3x^4 + x^3 - 2x^2$
3. $28x^2 - 23x - 15$
4. $-12a^2 + 4ab + 21b^2$
5. $16c^2 + 18cd - 9d^2$
6. $9y^2 + 6yz + z^2$
7. $-2c^2 + 9cd - 5d^2$
8. $-21a^2 + 34a - 8$

9. $24e^2 - 14ef - 3f^2$
10. $-24x^2 - 41xy - 12y^2$
11. $10x^2 + 21xy - 10y^2$
12. $10a^2 + 11ab - 6b^2$
13. $ab + 4ax - 2b^2 - 8bx$
14. $20a^2b^2 - abc - c^2$
15. $-3x^2 - 7xy - 2y^2$
16. $-35m^2 + 31mn - 6n^2$

17. $9x^2 - 16y^2$
18. $3 + 2p - 8p^2$
19. $10a^2 - 13ab - 30b^2$
20. $49 - 70mp + 25m^2p^2$
21. $10x^2 - 19x - 15$
22. $3c^2 - 5cd - 2d^2$
23. $-2a^2 - 7ab - 3b^2$

24. $10r^2 + 16rs + 6s^2$
25. $16a^2 + 40ac + 25c^2$
26. $e^2 - f^2$
27. $-2c^2 + 3cd + 9d^2$
28. $-20x^2 - 9x + 18$
29. $42x^2 - 29xy - 5y^2$
30. $9x^2 - 42xy + 49y^2$

LESSON 33

1. $9x^2 - 15x + 4$
2. $20x^2 + 17x - 63$
3. $x^3 + 3x^2 + 6x + 8$
4. $3x^3 + 16x^2 + 8x + 15$

5. $2x^3 + 6x^2 + 9x + 10$
6. $4x^3 + 13x^2 - 6x + 24$
7. $x^3 + 7x^2 + 15x + 25$

8. $a^4 + 5a^3 + 4a^2 - 7a - 28$
9. $b^3 + 2b^2 - 11b - 24$
10. $16c^2 - 14c - 15$

LESSON 34

1. $4x^3 + 2x^2 - x - 6$

2. $4x^4 + 3x^3 + 2x^2 + x + \dfrac{1}{2}$

3. $4x^3 - 3x^2 + x - 2$

4. $x^3 + 4x^2 - \dfrac{9}{4}x + 3$

5. $5x^3 + 7x^2 - 3x + 2$

6. $5x^4 + 4x^3 - 6x^2 + 3x - 1$

7. $4x^2 + 2x + 5$

8. $4a^3 - 3a^2 + 5a + \dfrac{5}{2}$

9. $5x^4 - \dfrac{15}{4}x^3 + 2x^2 + \dfrac{3}{2}a - \dfrac{3}{4}$

10. $2ab - 6 + \dfrac{4}{ab}$

11. $3z^4 - 6z^3 + 4z^2 - 2z + 1$

12. $3x^3 + 4x^2 + 2x - 1 + \dfrac{2}{3x}$

13. $-2x^2 + 5x - 3$

14. $2x^2y - 3xz + \dfrac{4}{y}$

15. $-4a^2 + 2a + 6$

16. $-5x^2 - 7x + 2$

17. $3s^2 - 9rs - 16r^2$

18. $\dfrac{-6x^3y}{z} - \dfrac{4z}{y} + \dfrac{2y^2z^3}{x^2}$

19. $5r - 2s$

20. $4x^2y - 3xy^2z + 2y^3z^2 - \dfrac{y^4z^3}{x}$

21. $12x^3 - 9x^2 + 27x - 3$

22. $4x^3 - 3x^2 - 6x + 10$

23. $3a^2b^2c^2 - 5abc + 2$

LESSON 35

1. $4x^2 + 11x + 21 + \frac{-2}{x-3}$

2. $-x^2 + x + 3 + \frac{3}{x+1}$

3. $z^2 + 9z + 81$

4. $-2x^2 + x - 2$

5. $3x^2 + 4x + 3 + \frac{3}{x+1}$

6. $2x^2 + 3x - 6$

7. $3x^2 - 7x - 1$

8. $12y^2 + 38y + 103 + \frac{312}{y-3}$

9. $a^2 + 4a + 16$

10. $x - 3 + \frac{-11}{2x-1}$

11. $4x^2 - 4x + 1$

12. $x^2 - 3x + 2$

LESSON 36

1. $-8x^3 - 12x^2 - 4$

2. $-8x^3 - 7x - 12$

3. $22x^4 + 26x^3 + 3x + 6$

4. $-6x^5 - 11x^4 - 7x^3 - 18x^2 - 28x - 7$

5. a

6. $25x^4 - 50x^2 + 25$

7. $6x^2 + 17x + 7$

8. $x^3 - 17x - 4$

9. a

10. $3x^2 + 5x + 8 + \frac{-4}{x-2}$

11. $x^2 - 2x - 1 + \frac{3}{x+2}$

12. $n^2 - 3n + 9$

13. $4x^2 - 2x - 2$

14. $6x^2 + 4x + 2 + \frac{1}{x-1}$

15. d

16. $-6x^5 + 7x^4 + 11x^3 - 15x^2 - 16x + 34$

17. $12x^5 + 14x^4 - 4x^3 - 9x^2 - 7x - 11$

18. $-4x^5 - 18x^4 + 16x^3 - 20x^2 - 16x - 22$

19. $-9x^5 - 9x^4 + 8x^3 + 34x^2 + 30x + 23$

20. d

21. $x^3 + 2x^2 - 5x + 12$

22. $x^3 + 6x^2 + 10x + 3$

23. a

24. $6x^2 + 12x - 48$

25. $8y^2 - 4y + 3 + \frac{-5}{y}$

LESSON 37

1. $2x^2(1 + 2x)$

2. $16(a^2 + 2b^2)$

3. $2(2x^2 + 3xy + 4y^2)$

4. $3x^2(x^2 + 4x + 6)$

5. $4x(x + 4x^2 + 8)$

6. $5x^2(1 + 3x + 5x^2)$

7. $9y^2(9y - 4x)$

8. $3a^2b^2(3ab + 1 + 2a^2b^2)$

9. $8y^3z^3(1 + 7y)$

10. $12a^2b^2c^3(5bc - 2a)$

11. $2xyz(9xyz + 10)$

12. $16(y^2 + 2y + 1)$

13. $4x^2y^2(3x^3y - 4)$

14. $4y(7y - 5 + 4y^2)$

15. $6x^2y^3z^2(8x^2 - z^2)$

16. $7(7x^2 - 9y^3)$

17. $7x^2y^3(1 + 3xy + 4x^2y^2)$

18. $9mp^2(6mnp + 3n^2 - 2m^2)$

19. $3ab(3a^2bc^2 - 6c^2 + 4a - 10bc^2)$

20. $4efg^2(4fg - 2ef + e - 5e^2fg)$

21. $7x^2y^2z^2(2xz^2 - 4y + 5x^3z - 7y^2z^3)$

22. $5a^2bc^2(3a + 5b - 6a^2b^2c^2 + 12abc^2)$

23. $6L^2m^2n^3(3mn - 6Lm + 4L^2m^2n^2 + 8L)$

24. $3r^2s^2t^2(7st^2 + 10r + 2s^2t^3 + 6r^2t^2)$

LESSON 38

1. $(a + 6)(a - 6)$

2. $4(x + 3)(x - 3)$

3. $a^2b^2(1 + ab)$

4. $(a - 4)(a^2 + 4a + 16)$

5. $9(x - 3y)(x + 3y)$

6. $(2b + 3)(4b^2 - 6b + 9)$

7. $(xy + a)(xy - a)$

8. $(4x - 5)(16x^2 + 20x + 25)$

9. $(4e + 7f)(4e - 7f)$

10. $4(2e + 3f)(2e - 3f)$

11. $(cd + 1)(c2d^2 - cd + 1)$

12. $(1 + 2a)(1 - 2a)$

13. $25(2x + 1)(2x - 1)$

14. $(1 + 2z)(1 - 2z + 4z^2)$

15. $9x^2(3x + a)(3x - a)$

16. $16(a + 2b)(a - 2b)$

17. $(y - 6x)(y^2 + 6xy + 36x^2)$

18. $(5x + 4y^2)(25x^2 - 20xy^2 + 16y^4)$

19. $(b + 3a)(b - 3a)$

20. $(7m^2 - 2n)(49m^4 + 14mn + 4n^2)$

21. $(1 + 4x)(1 - 4x + 16x^2)$

22. $64(y^2 - 2z)(y^4 + 2y^2z + 4z^2)$

23. $(7y + 2z)(7y - 2z)$

24. $(2a + 5)(4a^2 - 10a + 25)$

25. $125(2 - p^2)(4 + 2p^2 + p^4)$

26. $16(3a + 2b)(3a - 2b)$

27. $(5c + 4d)(25c^2 - 20cd + 16d^2)$

28. $(6x^2 - 7y^3)(36x^4 + 42x^2y^3 + 49y^6)$

29. $(2x - 1)(2x + 1)(4x^2 + 2x + 1)(4x^2 - 2x + 1)$

30. $(1 + 13x^2)(1 - 13x^2)$

LESSON 39
1. $(2 + c)(a + b + x)$
2. $(y + 2)(x2 - 5)$
3. $(x - 2)(x - 6)$
4. $(p + 5)(q + 2)$
5. $(2x + 3)(y + 1)$
6. $(a - 2)(2a + 3b)$
7. $(x + 3)(x2 - 5)$
8. $(a + 4)(6x + 1)$
9. $(a - 5)(x2 + y2)$
10. $(x - 5)(y - 3)$ or $(5 - x)(3 - y)$

11. $(x + 3)(y + 2)$
12. $(a + bc)(ab + c)$
13. $(a - 7)(a + 2b)$
14. $(x + 2)(x - y)$
15. $(a + b)(p - x)$
16. $3x(y - 4x)(x + 1)$
17. $5y(2x - 1)(y + 3)$
18. $b2(a + b)(b + 6)$
19. $4(2 - y^2)(x + 1)$

20. $(y - 5)(x^2 + r)$
21. $(y + 2)(y^2 - 3)$
22. $(3a - 1)(3b + 4)$
23. $(x + 1)(x^2 + 9)$
24. $(c - 4)(2c - 5d)$
25. $(2x - 3)(y - 4)$
26. $(3a + 2b)(6a - c)$
27. $(x + 3y)(5x - 2z)$
28. $(x - 2)(y + 1)(y - 1)$

LESSON 40
1. $(4x + 3y)(5x - 2y)$
2. $(4x + 5)(2x - 5)$
3. $(s + 4)(s + 7)$
4. $(4x - 3y)(2x - 5y)$

5. $(2x + 3y^2)(4x^2 - 6xy^2 + 9y^4)$
6. $(4x + 5y)(4x - 5y)$
7. $x^2(y + 2z)(y^2 - 2yz + 4z^2)$
8. $(2 + x)(2 - x)(16 + 4x^2 + x^4)$

9. $(x + y)(x - y)(1 + z)(1 - z)$
10. $x(4 + x)(4 - x)$
11. $25(x^2 + 3y^3)(x^2 - 3y^3)$
12. $(4a^2 - b)(4a - b^2)$

LESSON 41
1. $(a + 3)(a - 5)$
2. $(x + 9)(x - 4)$
3. $(b - 3)(b + 7)$
4. $(x - 7)(x + 5)$

5. $(x - 2)(x + 13)$
6. $(c + 8)(c - 5)$
7. $(r + 12)(r - 8)$
8. $(m + 9)(m - 6)$

9. $(y - 4)(y +7)$
10. $(e + 9)(e - 11)$
11. $(f + 16)(f - 5)$
12. $(a - 12)(a + 3)$

13. $(x - 8)(x + 6)$
14. $(x - 10)(x + 5)$
15. $(x + 12)(x - 5)$
16. $(e - 8)(e - 9)$

LESSON 42
1. $(7x + 2)(x + 3)$
2. $(2x - 7)(x - 2)$
3. $(4x - 3)(x + 5)$
4. $(5x - 2)(x + 3)$
5. $(3x + 1)(x + 4)$

6. $(5x - 4)(x + 2)$
7. $(4y - 3)(y - 2)$
8. $(3x - 5)(3x - 2)$
9. $(3x + 4)(2x - 1)$
10. $(2n - 5)(3n + 2)$

11. $(4x - 7)(x + 1)$
12. $(3a - 2)(3a - 4)$
13. $(5x - 4)(2x - 3)$
14. $(3x - 1)(2x + 1)$
15. $(x + 2)(2x + 3)$

16. $(3x - 1)(4x + 5)$
17. $(5x - 1)(4x + 3)$
18. $(5x + 2)(3x - 1)$
19. $(3x - 4)(2x - 3)$
20. $(2b - 7)(b + 4)$

LESSON 43

1 (a) $6(x + 2)(x - 3)$
  (b) $4x^3(x + 6)(x - 1)$
2. 1 and 11
3. The factor of 9 should be factored out.
4. Answer will vary. Find two numbers whose sum is -9 and whose product is 20.
5. $(x + 7)(x - 8)$
6. 3 and 8
7 (a) $(x + 12)(x - 10)$
  (b) $3x(x - 2y)(x + 5y)$
  (c) $2(x - 2)(x - 3)$
8. The numbers "a" and "b" must have the same sign (both positive or both negative).
9 (a) $(x + 11y)(x - 9y)$
  (b) $5x(x - 3)(x + 5)$
  (c) prime
10. The factorization yields a middle term of -2x rather than +2x. The correct factorization is $(x + 5)(x - 3)$.

11 (a) $(x + 4)(x - 12)$
   (b) $(x + 15)(x - 1)$
12 (a) $(x - 4)$
   (b) $(x + 5)$
13. $(x + 5y)(x - 2y)$
14. c
15. b
16. a
17. d
18. b
19. $(x + 8)(3x + 1)$
20. $(3h - 4)(3h - 4)$
21. $2a^4(a - 3)(a + 7)$
22. $(3x - 2)(5x - 3)$
23. missing the "3k"
24. $(7 + a)(1 + a)$

LESSON 44

1. $(u + 6v)(u - 9v)$
2. $(u + 2v)(u - 7v)$
3. $(st - 13)(st - 2)$
4. $(xy + 3)(xy + 4)$
5. $(10 - m)(m + 4)$
   or $(m - 10)(-m - 4)$
6. $(x + 7y)(x - 4y)$
7. $(x + 6)(x - 7)$
8. $(x + 16y)(x - 12y)$
9. $(u + 5v)(u - 9v)$
10. $(u + 3v)(u - 5v)$

11. $(3x + 2)(4x + 3)$
12. $(3y + 4)(5y + 2)$
13. $(3z - 4)(4z + 3)$
14. $(2z + 3)(4z - 3)$
15. $(8mn - 5)(3mn - 8)$
16. $(4m + 7n)(7m + 8n)$
17. $(3x + 4)(5x + 2)$
18. $(2y + 3)(3y + 4)$
19. $(4z - 3)(3z + 4)$
20. $(4z + 3)(5z - 2)$

21. $(4 - 3y)(5 + 2y)$
22. $(7y - 4z)(49y^2 + 28yz + 16z^2)$
23. $(3 + a)(2 - b)$
24. $(-x + 5)(x + 6)$
25. $(x - 3)(3x + 4)$
26. $(y^2 + 8x^2)(y^2 - 8x^2)$
27. $(-x + 2)(x - 3)$
28. $(1 + 4x^2)(1 - 4x^2 + 16x^4)$
29. $5y(x^2 - y)(2y + 3)$
30. $(x + 2)(y - 1)(y + 1)(x^2 - 2x + 4)$

LESSON 45

1. $\{2,5\}$
2. $\{-4,-2\}$
3. $\{-2,3\}$
4. $\{0,8\}$
5. $\{2,5/2\}$
6. $\{-1,4/3\}$
7. $\{-4,6\}$
8. $\{-3,5/2\}$
9. $\{-5,5\}$
10. $\{-3,3\}$
11. $\{1/2,2\}$
12. $\{-2,-1/2\}$
13. $\{-1,6\}$
14. $\{2/5,2\}$

LESSON 46

1. $2\sqrt{5}$
2. $3\sqrt{6}$
3. 7
4. $27\sqrt{2}$
5. $4\sqrt{10}$
6. $10\sqrt{2}$
7. $\frac{5}{3}\sqrt{2}$
8. $5\sqrt{6}$
9. 20
10. $6\sqrt{7}$
11. $4\sqrt{6}$
12. $12\sqrt{5}$
13. $2\sqrt{3}$
14. $20\sqrt{3}$
15. 12
16. $7\sqrt{2}$
17. $8\sqrt{7}$
18. $12\sqrt{2}$

LESSON 47

1. $-1 + \sqrt{6}, -1 - \sqrt{6}$
2. $-3 + \sqrt{14}, -3 - \sqrt{14}$
3. $-\frac{1}{2} + \frac{\sqrt{13}}{2}, -\frac{1}{2} - \frac{\sqrt{13}}{2}$
4. $\frac{5}{2} + \frac{\sqrt{21}}{2}, \frac{5}{2} - \frac{\sqrt{21}}{2}$
5. $-2 + \sqrt{2}, -2 - \sqrt{2}$
6. $-2 + \sqrt{7}, -2 - \sqrt{7}$
7. $-1/2, 2$
8. $-7, 3$
9. $-8, 6$
10. $1 + \sqrt{6}, 1 - \sqrt{6}$
11. $\frac{5}{2} + \frac{3\sqrt{3}}{2}, \frac{5}{2} - \frac{3\sqrt{3}}{2}$
12. $-2, 8$

LESSON 48

1. $\{5 \pm 2\sqrt{10}\}$
2. $\left\{\frac{5 \pm \sqrt{73}}{6}\right\}$
3. $\left\{\frac{1 \pm \sqrt{19}}{3}\right\}$
4. $\{-1 \pm \sqrt{5}\}$
5. $\left\{\frac{3 \pm \sqrt{21}}{2}\right\}$
6. $\{-6, 1\}$
7. $\{-3 \pm \sqrt{13}\}$
8. $\{1 \pm \sqrt{6}\}$
9. $\{-2, 6\}$
10. $\{-7, 2\}$
11. $\left\{\frac{-7 \pm \sqrt{41}}{4}\right\}$
12. $\left\{\frac{-5 \pm \sqrt{13}}{6}\right\}$
13. $\left\{\frac{5 \pm \sqrt{17}}{4}\right\}$
14. $\left\{\frac{-9 \pm \sqrt{129}}{2}\right\}$
15. $\{10 \pm \sqrt{110}\}$
16. $\left\{2 \pm \frac{\sqrt{14}}{6}\right\}$
17. $\{-2 \pm 2\sqrt{5}\}$
18. $\left\{\frac{1 \pm \sqrt{7}}{2}\right\}$
19. $\left\{\frac{5 \pm 3\sqrt{3}}{2}\right\}$
20. $\left\{\frac{-5 \pm \sqrt{33}}{2}\right\}$

LESSON 49

1. $\{4\}$
2. $\{1, -2/3\}$
3. $\left\{\frac{2 \pm \sqrt{14}}{5}\right\}$
4. $\{-2/3, 2\}$
5. $\{-2 \pm \sqrt{6}\}$
6. $\left\{\frac{5 \pm \sqrt{43}}{2}\right\}$
7. $\{2, 6\}$
8. $\{-3, 5/2\}$
9. $\{-1, 2/5\}$
10. $\left\{\frac{5 \pm \sqrt{41}}{2}\right\}$
11. $\{-1, 1/3\}$
12. $\{-2, 8\}$
13. $\left\{1 \pm \frac{\sqrt{6}}{2}\right\}$
14. $\{-8, 6\}$
15. $\{1 \pm \sqrt{5}\}$
16. $\left\{\frac{3 \pm \sqrt{137}}{8}\right\}$

LESSON 50

1. x = 1/3, y = 1/2
2. a = 5, b = 2
3. x = 5, y = 6
4. m = 3, n = 6
5. a = 1, b = -1
6. x = 4, y = 3
7. a = -1, b = -4
8. a = 8, b = 12
9. e = 4, f = 5
10. a = 500, b = 300
11. a = 1, b = 1
12. a = 11/4, b = 1/8

LESSON 51

1. a = 3, b = 2
2. x = 3, y = -3
3. c = 3, d = 5
4. e = -2, f = 3
5. a = 1, b = -4
6. x = -1, z = -1
7. a = 1/3, b = 0
8. p = 6, r = 2
9. m = 8, n = 6
10. x = -1, y = 3
11. x = 8, y = 5
12. a = 3, b = 1

LESSON 52

1. {3,2}   **2.** {4,-1}   **3.** {6,3}   **4.** {0,-4}   **5.** {29/7,-2/7}   **6.** {1,3}   **7.** {3,2/3}   **8.** {4,-3}

LESSON 53

1. {8/5, -19/5}
2. {2, 0}
3. {25/11, -2/11}
4. {1, 1}
5. {1/2, -1}
6. {-13/4, -11/12}
7. {2, -4}
8. {5/3, 2}
9. {-3/2, -1/4}
10. {3, 2}
11. {2, 3}
12. {7, 2}

LESSON 54

1. 2 1/2 hrs
2. 1:40 PM
3. 30 miles
4. 16 km/h, 32 km/h
5. 600 miles
6. 66 2/3 miles
7. 2 hrs
8. 42 mph
9. 10:45 AM
10. 11:34 AM
11. passenger: 88 mph, freight: 22 mph
12. slower plane: 230 mph, 690 miles; faster plane: 270 mph, 810 miles
13. 65.45 miles
14. 30 mph and 50 mph

LESSON 55

1. 25 quarters, 35 dimes
2. 22 nickels, 28 dimes
3. $3.45
4. 15 dimes
5. 3 quarters, 14 dimes
6. 75 coins
7. $21.55
8. 37 nickels
9. $14
10. 8 quarters, 20 dimes

LESSON 56

1. 47   **2.** 57   **3.** 85   **4.** 93   **5.** 408   6. 28   **7.** 24   **8.** 365   **9.** 81   **10.** 63

LESSON 57

1. peas: 18¢, beans: 8¢
2. $100
3. 12 heifers
4. 144 cans
5. carnation: 25¢, rose: 15¢
6. 5 men
7. pen: 30¢, pencil: 55¢
8. fixed cost: 25¢, extra word: 4¢
9. 50 seniors
10. soup: 99¢, spaghetti: $1.20
11. 22 shares for $24
12. $240
13. beans: 65¢, soup: 99¢
14. $200

LESSON 58

1. Wally: 30, Beaver: 10
2. man: 36, daughter: 14
3. Morgan: 36, Abbie: 20
4. Mickey: 23 1/2, Matt: 4 1/2
5. 8 years
6. 15 years
7. Sally: 9, Christine: 36
8. Mrs. Hall: 40, Mrs. Ryan: 25
9. 15 years
10. father: 40, son: 10

LESSON 59

1. 20 g
2. 18 3/4 lbs of buckwheat, 31 1/4 lbs of rolled oats
3. 50.4 oz
4. 50 lbs of 6%, 25 lbs of 10%
5. 24 lbs of 60¢, 16 lbs of 85¢
6. 54 gal
7. 5 1/3 qt of 35%, 10 2/3 qt of 50%
8. 9 lbs
9. 20 qt
10. 30%
11. 1.5 g
12. 40 g of each
13. 168 lbs of 26%, 112 lbs of 36%
14. 1 1/5 qt
15. 6 lbs chocolate, 4 lbs caramels

LESSON 60
1. 1 5/7 hrs
2. 1 3/7 hrs
3. 11 1/4 hrs
4. 2 2/5 hrs
5. 5 hrs
6. 26 2/3 min
7. 12 hrs
8. 1 7/8 hrs
9. 12 min
10. larger pipe: 8 hrs, smaller pipe: 24 hrs
11. 2 hrs
12. Henry: 12 days, Clyde: 6 days

LESSON 61
1. $270
2. 4 yrs
3. $550 at 5 1/2%, $1087.50 at 7 1/4%,$1637.50
4. 5% quarterly
5. $14
6. $20,000 in each
7. $1000
8. $45, $47.03, $1092.03
9. $8520 at 8%, $5680 at 10%
10. $15,000 at 9%,$30,000 at 8%
11. $2400 at 6%,$3600 at 4%
12. $14000 at 7%,$7000 at 9%, $21,000

LESSON 62
1. 9, 16
2. 6, 8, 10
3. 12, 21
4. 10
5. 8/13
6. 14,16,18
7. 12, 24
8. 4, 12
9. 4, 9
10. 11, 47
11. 36, 8
12. 4

LESSON 63
1. 34, 35
2. 9
3. 8
4. Macy:9,Sylvester:12
5. 95
6.Tom:$170,Amy:$230
7. -5, -3, -1, 1
8. 24
9. 12,13,14,15,16,17
10. 4 hrs

LESSON 64
1. 20%
2. $2.16
3. $55
4. $143.65
5. $60
6. $140
7. $163.71
8. $2500
9. 200
10. 4 min
11. 600
12. $48
13. 180
14. $5
15. 80
16. 20%
17. 12%
18. $28.50
19. $7215
20. $60
21. $78
22. $70
23. 24

LESSON 65
1. 18 ft by 30 ft
2. 10 in. by 26 in.
3. 20 ft by 30 ft
4. height: 12 ft, base: 6 ft
5. height: 4 ft, base: 6 ft
6. 3 inches
7. 4
8. 8 ft
9. 10 ft
10. 30 ft by 60 ft
11. 4 in.
12. 12 in. by 4 in. or 8 in. by 6 in.
13. small: 4 in., large: 8 in.
14. 20 in., 24 in., 24 in.
15. L: 9 in., W: 6 in.
16. L: 9 in., W: 8 in. or L: 8 in., W: 9 in.
17. L:11 in, W: 5 in.
18. L: 31 in., W: 12 in.
19. 15 in., 8 in.
20. 12 ft, 5 ft

LESSON 66
1. 4.5 hr
2. 140 quarters, 420 nickels
3. 28
4. corn: 50¢, beans: 60¢
5. 6 yrs
6. $2.46
7. $33.75
8. Lily, $10
9. 4
10. 92
11. $89.85
12. 7 ft by 5 ft or 5ft by 7 ft

LESSON 67
1. (a) yes
   (b) no, 2 has 5
       different values
   (c) no, 5 has 2
       different values
   (d) yes
   (e) yes
   (f) yes
2. (a) yes
   (b) yes
2. (c) no
3. (a) (i) 2
       (ii) -1
       (iii) -7
   (b) (i) 18
       (ii) 13
       (iii) 9
3. (c) (i) 9
       (ii) 21
       (iii) 41
3. (d) (i) 3
       (ii) -24
       (iii) -9
   (e) (i) -7
       (ii) 1
       (iii) -4
4. (a) (i) 4
       (ii) -7/4
       (iii) 5/2
       (iv) 6
4. (b) (i) 16
       (ii) 57/16
       (iii) 7/4
       (iv) 12
   (c) (i) 40
       (ii) 9/4
       (iii) 9/4
       (iv) 16
4. (d) (i) 44
       (ii) 139/16
       (iii) 34/19
       (iv) 12
   (e) (i) 26
       (ii) 43/8
       (iii) 20/11
       (iv) 6

LESSON 68

1. 2   **4.** $4b^3 - 23b^2 + 27b - 9$ **7.** −16   **10.** −18   **13.** −13   **15.** 207
2. -8   **5.** 39   **8.** 3/5   **11.** −13   **14.** 87   **16.** 36
3. $a^2 - a$ **6.** $4x^2 - 20x + 9$ **9.** 303   **12.** 14

17. $y = |x + 1|$ is a function

x	-2	-1	0	1	2
y	1	0	1	2	3

18. $y = -|x|$ is a function

x	-2	-1	0	1	2
y	-2	-1	0	-1	-2

19. $x = |y + 2|$ is not a function

x	0	1	2	3	4
y	-2	-1	0	1	2

20. $y = |x + 3|$ is a function

x	-2	-1	0	1	2
y	1	2	3	4	5

LESSON 69

1. $y = 4x^2$, domain: all real numbers, range: $y \geq 0$, vertex: (0, 0), minimum point

x	-2	-1	0	1	2
y	16	4	0	4	16

2. $y = -2x^2$, domain: all real numbers, range: $y \leq 0$, vertex: (0, 0), maximum point

x	-2	-1	0	1	2
y	-8	-2	0	-2	-8

3. $y = -\frac{1}{2}x^2$, domain: all real numbers range: $y \geq 0$, vertex: (0, 0), maximum point

x	-2	-1	0	1	2
y	-2	$-\frac{1}{2}$	0	$-\frac{1}{2}$	-2

2. $y = -3x^2$, domain: all real numbers, range: $y \leq 0$, vertex: (0, 0), maximum point

x	-2	-1	0	1	2
y	-12	-3	0	-3	-12

5. $y = 2x^2$, domain: all real numbers, range: $y \geq 0$, vertex: (0, 0), minimum point

x	-2	-1	0	1	2
y	8	2	0	2	8

6. $y = \frac{1}{4}x^2$, domain: all real numbers, range: $y \geq 0$, vertex: (0, 0), minimum point

x	-2	-1	0	1	2
y	1	$\frac{1}{4}$	0	$\frac{1}{4}$	1

LESSON 70

1. $y = -x^2 - 2$, domain: all real numbers, range: $y \leq -2$, vertex: (0, -2), maximum point

x	-2	-1	0	1	2
y	-6	-3	-2	-3	-6

2. $y = \frac{1}{2}x^2 + 2$, domain: all real numbers, range: $y \geq 2$, vertex: (0, 2), minimum point

x	-2	-1	0	1	2
y	4	5/2	2	5/2	4

3. $y = -x^2 + 4$, domain: all real numbers, range: $y \leq 4$, vertex: (0, 4), maximum point

x	-2	-1	0	1	2
y	0	3	4	3	0

4. $y = -x^2 + 3$, domain: all real numbers, range: $y \leq 3$, vertex (0, 3), maximum point

x	-2	-1	0	1	2
y	-1	2	3	2	-1

240

5. $y = x^2 - 1$, domain: all real numbers,
range: $y \geq -1$, vertex: (0, -1), minimum point

x	-2	-1	0	1	2
y	3	0	-1	0	3

6. $y = -\frac{1}{2}x^2 - 4$, domain: all real numbers,
range: $y \leq -4$, vertex: (0, -4), maximum point

x	-2	-1	0	1	2
y	-6	$-4\frac{1}{2}$	-4	$-4\frac{1}{2}$	-6

7. $y = x^2 + 1$, domain: all real numbers,
range: $y \geq 1$, vertex: (0, 1), minimum point

x	-2	-1	0	1	2
y	5	2	1	2	5

8. $y = \frac{1}{4}x^2 - 1$, domain: all real numbers,
range: $y \geq -1$, vertex: (0, -1), minimum point

x	-2	-1	0	1	2
y	0	$-\frac{3}{4}$	-1	$-\frac{3}{4}$	0

LESSON 71

1. $y = (x - 1)^2 + 3$, domain: all real numbers,
range: $y \geq 3$, vertex: (1, 3), minimum point
axis of symmetry: x = 1

x	-2	-1	0	1	2	3
y	12	7	4	3	4	7

2. $y = -(x + 2)^2 + 2$, domain: all real numbers
range: $y \leq 2$, vertex: (-2, 2), maximum point,
axis of symmetry: x = -2

x	-4	-3	-2	-1	0	1
y	-2	1	2	1	-2	-7

3. $y = -(x + 1)^2 - 1$, domain: all real numbers,
range: $y \leq -1$, vertex: (-1, -1), maximum point
axis of symmetry: x = -1

x	-4	-3	-2	-1	0	1
y	-10	-5	-2	-1	-2	-5

4. $y = (x - 3)^2 + 4$, domain: all real numbers,
range: $y \geq 4$, vertex (3, 4), minimum point,
axis of symmetry: x = 3

x	0	1	2	3	4	5	6
y	13	8	5	4	5	8	13

5. $y = 2(x - 1)^2 + 5$, domain: all real numbers,
range: $y \geq 5$, vertex: (1, 5), minimum point,
axis of symmetry: x = 1

x	-2	-1	0	1	2	3	4
y	23	13	7	5	7	13	23

6. $y = -2(x + 1)^2 - 5$, domain, all real numbers,
range: $y \leq -5$, vertex: (-1, -5), maximum point,
axis of symmetry: x = -1

x	-4	-3	-2	-1	0	1	2
y	-23	-13	-7	5	-7	-13	-23

7. $y = -(x + 1)^2 - 2$, domain: all real numbers,
range: $y \leq -2$, vertex: (-1, -2), maximum point,
axis of symmetry: x = -1

x	-4	-3	-2	-1	0	1	2
y	-11	-6	-3	-2	-3	-6	-11

8. $y = -(x - 1)^2 - 1$, domain, all real numbers,
range: $y \leq -1$, vertex: (1, -1), maximum point,
axis of symmetry: x = 1

x	-2	-1	0	1	2	3
y	-10	-5	-2	-1	-2	-5

9. $y = 2(x + 3)^2 - 1$, domain: all real numbers, range: $y \geq -1$, vertex: $(-3, -1)$, minimum point, axis of symmetry: $x = -3$

x	-5	-4	-3	-2	-1	0
y	7	1	-1	1	7	17

10. $y = (x + 3)^2 + 2$, domain: all real numbers, range: $y \geq 2$, vertex: $(-3, 2)$, minimum point, axis of symmetry: $x = -3$

x	-6	-5	-4	-3	-2	-1	0
y	11	6	3	2	3	6	11

LESSON 72
1. 0, rational
2. -11, irrational
3. 0, rational
4. 0, rational

5. 4, rational
6. 0, rational
7. 121, rational
8. 81, rational

9. 33, irrational
10. 100, rational
11. -23, irrational
12. 25, rational

13. 21, irrational
14. -12, irrational
15. 17, irrational
16. 1, rational

LESSON 73
1. 15 in., 20 in.
2. 8 in., 15 in.
3. 9 in., 12 in., 15 in.
4. 21 cm, 28 cm
5. 5 in., 12 in.
6. 12 cm, 16 cm
7. 36 cm
8. 7 in., 24 in.
9. 15 cm, 20 cm
10. 252 sq in.

LESSON 74
1. (a) $\sqrt{74}$
   (b) $\sqrt{162} = 9\sqrt{2}$
   (c) $\sqrt{101}$
   (d) $\sqrt{136} = 2\sqrt{34}$
  (e) $\sqrt{149}$
   (f) $3\sqrt{13}$
   (g) $\sqrt{290}$

1. (h) $\sqrt{58}$
2. $\sqrt{41} + \sqrt{85} + \sqrt{74}$
3. $14\sqrt{2}$
4. $H = (1, -4)$, $D = \sqrt{97}$
   $P = 26$, $A = 36$
5. $R = \sqrt{41}$
6. $R = 2\sqrt{5}$

7. $D = (1, 7)$, diagonal:
   $2\sqrt{10}$, $A = 20$
8. $P = 5 + 7\sqrt{5}$
9. $A = 10$
10. $A = 48$,
   $P = 4\sqrt{13} + 20$

LESSON 75
1. $x = 66$
2. $a = 49$
3. $b = 36$
4. $x = 12$
5. $y = 4$

6. no solution
7. $x = 4$
8. $a = 9$
9. $x = 9$
10. $x = 15$

11. no solution
12. $a = 11$
13. $x = 171$
14. $x = 144$
15. $x = 16/9$

16. $a = 437$
17. no solution
18. $x = 16$
19. $a = 46$
20. $y = 7, y = -7$

21. $y = 1$ only
22. $a = 3$
23. no solution
24. $x = -1$ only
25. $b = -8$

LESSON 76
1. (a) 64
  (b) 28
  (c) 30
  (d) 6
  (e) 36
  (f) 32

1. (g) $3/2 = 1.5$
  (h) 80
2. (a) 20
  (b) 32
  (c) 80/3
  (d) 16

2. (e) 5
  (f) 1
  (g) 128
  (h) 640
3. (a) direct
  (b) inverse

3. (c) inverse
  (d) direct
4. 3.5
5. 4/3
6. $A = kbh$, $k = 1/2$
7. $P = ks$, $k = 6$

8. $a = 160$, $b = 5$,
   $c = 280$, $d = 9$
9. (a) $xy = 40$
  (b) $RT = 100$
  (c) $CD = 144$

75-minute Review

1. $12\frac{53}{60}$

2. 3

3. $\frac{1}{4}$

4. $21\frac{17}{25}$

5. 283.192

6. $\frac{12}{7} = 1\frac{5}{7}$

7. $2.34

8. $6.10

9. x = 110

10. 37

11. 16 cookies

12. 40%

13. 78%

14. $7\frac{17}{18}$

15. 19

16. x = 30

17. 16

18. x = 6

19. $23\frac{1}{45}$

20. exponent: 3,
variable: y
coefficient: 17

21. -2

22. -6a + 9b – 12c + 15d

23. 12

24. 500

25. $40

26. 12

27. -12

28. 5

29. -4

30. 9x – 3y – 13z

31. 120

32. 2•2•2•3•3

33. x = 16

34. -5

35. $10\frac{71}{120}$

36. 29

37. 60 mph

38. 115

39. 3

40. $1\frac{1}{14}$

41. 0.35, 35%

42. 21

43. $\frac{3}{5}$

44. $x = 7\frac{29}{30}$

45. $x = 6\frac{53}{62}$

46. x = 6

47. -4

48. 5 gallons

49. 0.085, $\frac{17}{200}$

50. $14\frac{45}{62}$

243

Printed in the United States
By Bookmasters